T0234120

Finnie's Notes on Fracture Mechanics

C.K.H. Dharan • B.S. Kang • Iain Finnie

Finnie's Notes on Fracture Mechanics

Fundamental and Practical Lessons

 Springer

C.K.H. Dharan
University of California at Berkeley
Berkeley, CA, USA

Iain Finnie
University of California at Berkeley
Berkeley, CA, USA

B.S. Kang
Dean, School of Engineering
Pusan National University
Beon-gil, Busan, Korea
Republic of (South Korea)

ISBN 978-1-4939-7970-7   ISBN 978-1-4939-2477-6   (eBook)
DOI 10.1007/978-1-4939-2477-6

Printed on acid-free paper

This Springer imprint is published by Springer Nature
The registered company is Springer Science+Business Media New York

Preface

Fracture mechanics is integral to the analysis and verification of safety-critical structures and is incorporated into many of the codes and standards that govern their certification and use. Given its importance today, it is difficult to believe that this was not the case just 60 years ago. Although the basic principle of fracture mechanics, that the strength of a material is related to the presence of potential flaws in it, was proposed by Griffith in 1920, it was only after a series of tragic failures and disasters that took many lives that fracture mechanics methodology was developed and, over the years, grown to reach the state of maturity it is today. Most engineering programs at universities and technical institutes offer courses in fracture mechanics, and there are a number of textbooks dealing with the subject. However, few deliver an insight into the field that Iain Finnie provided in the notes that accompanied his lectures at the University of California, Berkeley. Over the years, following his retirement from teaching in 1999, the notes were no longer available and were in danger of disappearing altogether. The editors of this volume, who attended his lectures as students, felt that the publication of the notes with their in-depth explanation of the fundamentals of fracture mechanics and discussions of the limitations and assumptions inherent in much of the theory would be a much needed contribution to the field.

Most textbooks tend to follow an approach in which many of the relations used in fracture mechanics theory are either simply given in finished form or, in the more mathematically oriented texts, derived in more detail. There is often limited explanation of the applicability of the theory, or in the assumptions inherent in it, or its limitations. In addition, how the predictions from fracture mechanics can be affected by material properties affected by processing are rarely treated.

Professor Finnie's notes provide a clear and concise explanation of the field, delving into the foundations of the theory and crossing with ease the boundary between mechanics and mechanical behavior of materials. His practical approach while grounded in the fundamentals illustrated to the student the limitations of the theory in the context of structures made from real materials, giving the reader an insight into the use of fracture mechanics in the design of safety-critical structures, particularly in understanding how mathematical theories work when applied to mechanical behavior. In editing these notes, the editors have endeavored to keep the tenor and spirit of Iain Finnie's writing, making only the few changes we felt were necessary to add clarity and reduce the chance for misinterpretation. We hope that this publication will provide to the student and practicing engineer alike a deeper insight into the field of fracture mechanics, in both theory and applications to engineering materials and structures.

Berkeley, CA, USA   C.K.H. Dharan
Busan, Republic of South Korea   B.S. Kang
Berkeley, CA, USA   Iain Finnie

Biography

In 1974, **Iain Finnie** introduced a new course in fracture mechanics at Berkeley, a course that soon became highly popular among engineering graduate students. The approach taken by Finnie was to provide a rigorous theoretical exposition of the field accompanied by his own observations and examples of some of the problems he had encountered in his research and consulting practice. In this manner, he could relate the relative importance of the parameters in fracture theory as well as the applicability of the theory to practice. As students of Professor Finnie we appreciated his clear lectures, his perfect chalkboard handwriting, and his sense of humor. It was much later that we realized that what Iain Finnie taught us were a sense of the field, a clear view of the landscape, and the pitfalls that one can encounter. His lectures covered the historical development of the field emphasizing the often constrained sets of assumptions inherent in the theory. Thus, the student learned to keep in mind the framework on which a particular theory was constructed when developing failure criteria for a specific set of practical conditions and materials. He created a set of notes that reflected this approach which included problems that were unique, practical, and illustrative. His many years of experience consulting in the petrochemical and aerospace industries provided the bases for most of these problems. After his retirement in 1993, his notes were no longer in circulation and were in danger of being lost forever. It was with this concern in mind that we decided to edit his notes for this book. We believe that these notes will form a classic and never outdated treatise on the fascinating field of fracture and fracture mechanics and provide a level of understanding to both students and practicing engineers that is not readily available in existing texts in the field.

We provide here a brief biography of Professor Finnie excerpted from the 2009 University of California, Berkeley, publication, *In Memoriam*.

Iain Finnie, professor emeritus of mechanical engineering at the University of California, Berkeley, passed away peacefully at home on December 19, 2009, from pneumonia and complications of Parkinson's disease, with his wife and daughters at his bedside. He was 81.

Iain Finnie, Professor of
Mechanical Engineering,
Emeritus UC Berkeley
1928–2009

A pioneer and world expert in engineering materials, Professor Finnie served as department Chain, pioneer and world expert in engineering materials. Professor Finnie served as department chairman from 1979 to 1987. Elected to the National Academy of Engineering (NAE) in 1979 and an honorary member of the American Society of Mechanical Engineers (ASME) in 1981. His textbook, Creep of Engineering Materials, written with William Heller, is regarded as a classic in its field. When he retired in 1993, Professor Finnie received the Berkeley Citation, the highest honor the campus awards.

Born to Scottish parents in Hong Kong on July 18, 1928, he was schooled at George Watson's College in Edinburgh, Scotland, where he lived with his aunt, his uncle, and his "wee cousin" (Dr. Gilbert Kennedy, a rugby player for Scotland, with whom he was as close as a brother. In 1940, when women and children were evacuated from Hong Kong, Iain joined his mother and sister and lived in Victoria, BC, and Banff, Alberta, Canada. His father, the director of Swire & Sons' Taikoo Dockyard, was captured and imprisoned on Christmas Day, 1941, and released early in 1945 to help rebuild Hong Kong.

Iain attended the University of Glasgow and received an unrestricted fellowship that he used to complete an M.S. and D.Sc. at the Massachusetts Institute of Technology (MIT). He wrote problem sets for the books of Professor Jaap Den Hartog, a world expert on mechanical vibrations, staying summers at the Den Hartog family island home on Lake Winnipesaukee, New Hampshire. The Den Hartogs considered him their third son. After teaching a summer session at MIT and facilitating classes for world-renowned probability expert Professor Wallodi Weibull, Iain moved to Emeryville, California, to work for Shell Oil Development Co. In 1961, he joined the faculty at UC Berkeley and became a tenured professor at age 34.

In 1965, as part of a UC team led by Professor Erich Thomsen and funded by a Rockefeller grant, Professor Finnie helped establish the engineering department at Universidad Católica in Santiago, Chile. In 1967, he received a Guggenheim Award, a rarity for engineers, to study brittle solids—research that took him everywhere from miles down a South African gold mine to rock drilling in

Switzerland. In 1974, he received an honorary D.Sc. degree from the University of Glasgow, where he wore the hat of sixteenth-century-reformer John Knox at the ceremony.

Professor Finnie married Joan Roth McCorkindale, a widow with two young daughters, Carrie and Katie, in 1969. The couple had a daughter, Shauna. They took two sabbatical leaves, in 1976 and 1987, to Lausanne, Switzerland, where he was Professeur Invité at the École Polytechnique Fédérale de Lausanne (EPFL).

In the course of his career, Professor Finnie served as faculty advisor to over 40 Ph.D. students, with whom he wrote more than 185 papers on subjects such as fatigue of metals, crack propagation, and erosion of materials. He held a strong record for recruiting some of Berkeley's top talent and encouraging minority students to pursue careers in engineering. Many of his recruits have become university presidents, deans, and chaired professors worldwide. As a result of his inspiration and motivation, two of his daughters hold graduate degrees in mechanical engineering and are successful career women in the field. He remained close to his students over the years, sharing with them his sparkling sense of humor, his passion for skiing, and his hospitality. At one of the many parties in his Berkeley home, his graduate students donned T-shirts labeled "Finnie's Flaws" (consistent with their research on fracture mechanics) and presented him with one entitled "Master Flaw." Joan later gave their wives sashes emblazoned with "Finnie's Flawless Females." When Professor Finnie retired in 1993, his students framed his T-shirt and "retired" it, as is sometimes done on a professional athlete's retirement.

As his former students, Professor Finnie will always be remembered for his fine mind, quick wit, and his superlative teaching skills.

Berkeley, CA, USA   C. K. H. Dharan
Busan, Republic of South Korea   B. S. Kang

General References on Fracture

1. Liebowitz H (editor). Fracture. Academic Press; 1968.
   Volume 1: Microscopic and macroscopic fundamentals
   Volume 2: Mathematical fundamentals
   Volume 3: Engineering fundamentals and environmental effects
   Volume 4: Engineering fracture design
   Volume 5: Fracture design of structures
   Volume 6: Fracture of metals
   Volume 7: Fracture of non-metals and composites

2. Knott JF. Fundamentals of fracture mechanics. Butterworths, England: Halsted Press; 1973, (also available in USA).
3. Broek D. Elementary engineering fracture mechanics. Leyden: Noordhoff; 1974.
4. Lawn BR, Wilshaw TR. Fracture of brittle solids. Cambridge University Press; 1975.
5. Hertzberg RW. Deformation and fracture mechanics of engineering materials. John Wiley; 1976.
6. Rolfe J, Barsom J. Fracture and fatigue control in structures. Prentice Hall; 1977.
7. Hellan K. Introduction to fracture mechanics. McGraw-Hill; 1984.
8. Kanninen MF, Popelar CH. Advanced fracture mechanics. Oxford University Press; 1985.
9. Parker AP. The mechanics of fracture and fatigue. Spon-Methuen; 1981.
10. Many ASTM Special Technical Publications starting with STP 381. Fracture toughness testing and its applications; 1964.

Other sources and journals

Proceedings of International Congresses of Fracture
Engineering Fracture Mechanics
J. of Materials Science
Int. J. of Fracture
J. Eng. Materials and Technology (ASME)
Metallurgical Transactions

Contents

Chapter 1
The Nature of Fracture

1.1 Introduction

Our objective in this introductory chapter will be to describe the general features of the fracture process and to introduce some of the terminology that will be used. After doing this, we will attempt to trace the route by which our present understanding of fracture developed. The historical aspects of our topic are of considerable interest because some of the early discoveries were forgotten or not used for long periods. As with any subject there were many false starts and competing theories, some of which later became reconciled. One could make the same statement about the literature of the present day. Perhaps some feeling for the historical development of fracture theories will help provide perspective in sifting and evaluating the vast amount of literature that continues to appear in the field of fracture.

In discussing *fracture*, which we understand as the total or partial separation of a part, it is important to distinguish the phases of *crack initiation* and subsequent *crack propagation*. The propagation phase may involve slow (stable) growth or fast (unstable) growth.

Crack initiation occurs in many materials in the manufacturing process. Examples are surface cracks in glass due to handling, internal cracks in polycrystalline ceramics due to anisotropic thermal contraction on cooling, and flaws in welded structures due to faulty fabrication or design. The cracking or decohesion of second-phase particles or inclusions during cooling or plastic deformation initiate ductile fracture. Cracks may be produced by plastic deformation in polycrystalline materials at grain boundaries or other obstacles, and irradiation or diffusion at high temperatures may lead to the growth of voids.

In the past a lot of attention was given to the processes by which microscopic cracks could be initiated. More recently, the engineering viewpoint has become one of accepting the fact that some types of cracks or flaws will inevitably be present in real structures. An attempt is made to minimize the size and severity of these defects by choice of material, construction techniques, and nondestructive

© Springer Science+Business Media New York 2016
C.K.H. Dharan et al., *Finnie's Notes on Fracture Mechanics*,
DOI 10.1007/978-1-4939-2477-6_1

inspection. As a result, the emphasis in experimental studies of most types of fracture has changed from testing smooth specimens to studying the behavior of specimens containing preexisting cracks.

Slow stable crack growth occurs when the crack propagation rate may be controlled by adjusting the applied loads or displacements. Thus, if the load is removed, the crack growth will cease. In fact, in some polymeric materials, the "cracks" may heal if temperatures are high enough. Fatigue, stress corrosion, and creep rupture are types of fracture which are characterized by prolonged periods of slow crack growth prior to final fast fracture. Usually in short-term tests on brittle solids, such as ceramics, very little slow crack growth occurs, while in thin sheets of a ductile metal such as aluminum, steadily increasing loads may lead to extended periods of stable crack growth. However, whether slow crack growth occurs in these short-time tests is a function not only of the material but also of the test configuration. The question of crack stability will be discussed in Chap. 4. For the present we point out that it is possible to devise experiments where slow stable crack growth occurs in a material such as glass while materials normally thought of as ductile may fail in a fast unstable manner with little slow crack growth.

Fast unstable growth arises, essentially, when the combination of material and loading conditions is such that crack growth continues to release more energy than the fracture process can absorb. The crack accelerates rapidly with the excess energy appearing as kinetic energy of the material around the crack. Under these conditions very high crack growth velocities may result; for example, values up to 2000 m/s have been reported for steels. If the structure is loaded by dead weights or by gas, the crack may propagate a large distance before appreciable unloading occurs. A dramatic illustration is provided by fractures in natural gas pipe lines which have propagated in some cases for several kilometers. In a hydrostatically loaded structure the pressure is released more rapidly, and crack arrest may occur after limited propagation. The fast unstable type of crack growth is certainly dramatic and has received the greatest amount of attention in fracture studies.

However, it is not an inevitable aspect of fracture for, as pointed out earlier, with certain configurations of specimens, it is possible to obtain only slow stable crack growth in materials which absorb little energy during fracture.

The terms *elastic* and *plastic* should be familiar, but since much of our discussion of fracture hinges on the extent to which one or the other type of deformation dominates, they will be discussed briefly.

Elastic deformation "is that which recovers upon unloading."

To be more precise, to exclude "anelastic" or delayed recovery, we should specify that it recovers essentially immediately, the time limits being set by the velocities of elastic wave propagation. Except in rubbery polymer, elastic strains are usually small, are linearly proportional to stress, and are independent of the path of loading. This means that separate stress or deformation solutions may be superimposed unless instability is involved or a change in the boundary conditions is produced by loading as in the contact stress problem. These aspects of elastic behavior result in a great simplification compared to plastic deformation. As a result, in treating fracture, elastic solutions are used as much as possible before involving other types of material behavior.

Plastic deformation "is that which remains upon unloading."

Since it is dependent on the path of loading and varies in a nonlinear manner with the applied stress, it is no longer possible to superimpose separate solutions. Certain simplifications such as ignoring the path dependence and treating plastic deformation as a "nonlinear elasticity" are sometimes convenient. They can be misleading if applied to situations in which unloading occurs, such as behind a growing crack. Approximate solutions in which elastic strains and strain hardening are ignored allow upper and lower limits to be set on the loads that can be carried by a structure.

We now turn to two terms that present perhaps the greatest difficulty in offering definitions. They are *brittle* and *ductile*. When applied to a smooth unnotched tension specimen, their definition is straightforward and familiar. In a material which behaves in an ideally brittle manner, only elastic deformation occurs prior to separation. Thus, the pieces resulting from fracture could be fitted together to restore the original geometry of the part. A more practical definition of brittle behavior in a tension test might be that in which the plastic strain preceding fracture is less than several percent. By contrast, in a tension test on an ideally ductile material, deformation continues until the necked cross section has drawn down to a point. This type of fracture is sometimes referred to as *rupture* and is observed in testing metals such as lead and indium. In practice, separation usually occurs before the cross section has been so greatly reduced. We define ductile behavior as that in which large plastic strains precede separation of the specimen. The concepts of ideally brittle or ideally ductile behavior are often useful and allow, for example, analysis to be developed for shaping operations for these two classes of materials. Sometimes we speak about *brittle materials* and *ductile materials*. While such descriptions might seem adequate to describe say glass and 18 Cr–8 Ni austenitic stainless steels, it is found that the behavior of a material may change if the dimensions of the part, the stress state, the loading rate, the temperature, the environment or combinations of these are altered. For example, glass may form a continuous chip as in a ductile metal when scratched with a diamond, and austenitic stainless steels may fracture in the elastic range when exposed to hot aqueous chloride solutions. The very heart of our subject is the fact that low-carbon steels which would be classified as ductile by a tension test may fail at low nominal strains when some of the factors just listed are varied. Alternatively, fabricating processes may produce profound changes in mechanical behavior. Low-carbon steels, which show considerable ductility in a conventional tension test, have been shown to behave in a brittle manner in the same test if first precompressed to a large enough strain at elevated temperature.

If we consider now the fracture of a large structure containing a flaw, our description of the failure as brittle or ductile may depend on the scale at which we examine the failed structure. Despite local stresses exceeding yield at the flaw, the structure may have failed at nominal stresses (i.e., those computed by ignoring the flaw) below yield. From a designer's point of view, this is a brittle fracture. However, as the crack propagates, the local stresses at its tip will be large. When the fracture surface is examined on a local or microscale, it may show evidence of extensive plastic deformation. In an attempt to eliminate this confusion in terminology, the terms *frangible* and *tough* have been applied to describe fracture but have found limited usage. Frangible behavior is used to describe failures at nominal stresses below the yield with very limited plastic deformation *prior* to fracture. Tough behavior is used to describe fracture which occurs at nominal stresses above the yield.

Fracture can also be described in terms of the macroscopic or microscopic details of the fracture surface. On a macroscopic scale, fracture surfaces as in Fig. 1.1 can be described by the percentage of square or slant fracture. What we have called square fracture is often referred to as flat, brittle, or cleavage. The last term is not appropriate on a macroscale, and the other two may be misleading if the micromechanisms of fracture have involved extensive plastic deformation. Terms used essentially synonymously with slant fracture are oblique, shear, tearing, and ductile. In fracture tests, such as shown in Fig. 1.2a, on high-strength steels and alloys of titanium or aluminum, there is a transition from square to slant fracture as the plate thickness is decreased. This is related to a transition in the stress field at the

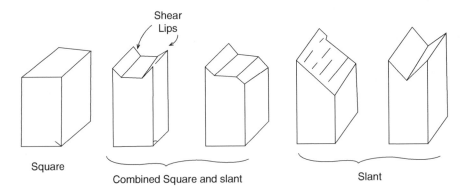

Fig. 1.1 Schematic of fracture surface appearance on a macroscopic scale ranging from 100 % square to 100 % slant. The slant fracture may form in either a symmetrical or asymmetrical form

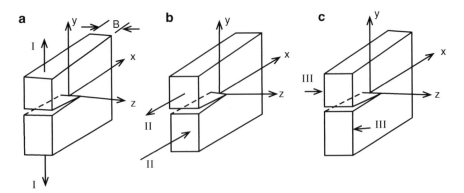

Fig. 1.2 Three models of loading a crack. (**a**) I direct opening, (**b**) II forward shear, and (**c**) III sideways shear (anti-plain strain)

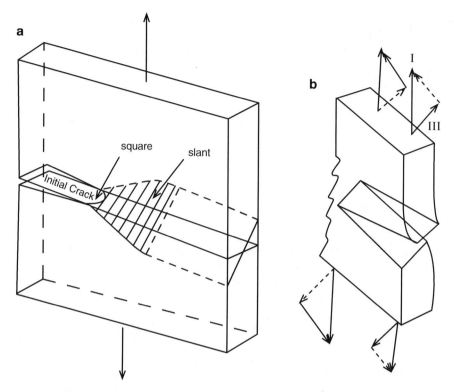

Fig. 1.3 (**a**) Development of slant fracture from a square initial crack. (**b**) Illustration that slant crack under loading shown is combination of Modes I and III

tip of the crack from *plane strain* to *plane stress*, concepts which will be discussed in Chap. 2. The manner in which a slant fracture surface develops from a square starter crack is sketched in Fig. 1.3a. Square fracture under the loading shown in Fig. 1.2a is referred to as Mode I or the direct opening mode. A slant fracture under the same loading is a combination of Mode I and Mode II loading as illustrated by Fig. 1.3b. In Chaps. 3 and 4, the stress distributions near the crack tip will be derived for elastic behavior and used to explain many aspects of fracture. Whether slant or square fracture is observed is also related to the microscopic processes involved in fracture. Before discussing these it is appropriate to consider the concept of the *ideal* strength or maximum possible strength of a perfect solid.

1.2 The Ideal Strength and Mechanisms of Failure

The ideal tensile strength is estimated by considering the stress required to separate two adjacent atomic planes. An exact calculation requires a detailed knowledge of the interatomic forces and has been carried out for special cases. However, for our

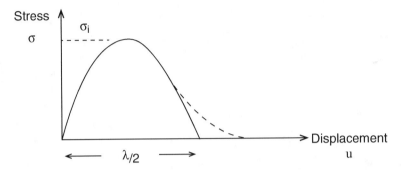

Fig. 1.4 Approximation used in derivation of theoretical strength

purposes an approximate calculation given by Orowan [1] is adequate. The half
sine wave of wavelength λ shown in Fig. 1.4 is taken as a rough approximation
to the stress-displacement curve for the separation of two atomic layers. That is
$\sigma = \sigma_i \sin(2\pi u/\lambda)$ where u is the amount by which the separation of two atomic
planes increases when stress is applied and σ_i is the maximum or ideal strength.

If the spacing of the atomic planes in the unstressed condition is b, then the strain
ε corresponding to displacement u is $\varepsilon = u/b$. For small values of $2\pi u/\lambda$ the stress-
displacement relation shown in the figure becomes $\sigma \simeq \sigma_i(2\pi u/\lambda) = \sigma_i(2\pi b/\lambda)\varepsilon$,
and since stress and strain are related through Young's Modulus E in the range of
linear elastic behavior,

$$\sigma_i \frac{2\pi b}{\lambda} = E \tag{1.1}$$

Another relationship is obtained from the fact that energy must be supplied to create
fresh surface. This surface energy arises because the bonding forces between atoms
at the surface are not symmetrical, and it can, in principle, be measured or estimated
in a number of ways. This surface energy, traditionally denoted by γ, is usually
expressed in ergs per cm^2 of each of the two surfaces formed by fracture and may be
thought of as a thermodynamic property of the material. Typical values quoted for γ
in ergs/cm^2 are 600 for silica glass, 1000 for aluminum oxide, 2000 for α iron, and
5400 for diamond.[1]

Returning now to Fig. 1.4 and assuming that all work done in separating the two
atomic layers appears as surface energy,

$$\gamma = \frac{1}{2}\int_0^{\lambda/2} \sigma_i \sin \frac{2\pi u}{\lambda} du = \frac{\sigma_i \lambda}{2\pi} \tag{1.2}$$

Combining Eqs. (1.1) and (1.2) leads to an estimate for the ideal strength:

---

[1] 1 erg cm^{-2} = 10^{-3} Jm^{-2} = 5.7 × 10^{-6} in lb/in^2.

$$\sigma_i = \left(\frac{E\lambda}{b}\right)^{1/2}$$

Taking γ and b for a variety of materials, the predictions are seen to fall in the range $E/10 < \sigma_i < E/5$. More accurate predictions, based on a detailed knowledge of atomic bonding, have been discussed by Kelly [2] and indicate that this simple approach may overestimate strength by a factor of about two in some cases. Thus, as a "rule of thumb,"

$$\frac{E}{10} < \sigma_i < \frac{E}{5} \quad \text{or} \quad \sigma_i \simeq \frac{E}{10} \tag{1.3}$$

Similar estimates for the ideal shear strength τ_i lead to the order of magnitude result $\tau_i \simeq G/10$ where G is the shear modulus.

These strength levels are far from realized when conventional bulk specimens are tested, although they are approached in whiskers of microscopic dimensions. Materials that would normally be classified as brittle based on a tension test, such as ceramics, typically fail at tensile stress levels of $E/1000$ rather than $E/10$ because materials have defects and flaws. Metals, which show ductile behavior in uniaxial tension tests, yield at shear stresses which are one or two orders of magnitude less than the ideal value.

These discrepancies between ideal and observed strength levels may be explained by noting that certain types of defects will inevitably be present in all but the most carefully prepared materials. In 1920, the elastic stress field around a sharp elliptical crack was used to explain the discrepancy between ideal and observed tensile strengths in solids showing brittle behavior. Later in 1934 the presence of line defects called *dislocations* in a crystalline lattice was shown to explain the discrepancy between ideal and observed shear strengths.

Returning to the microscopic mechanisms of fracture, two possibilities exist for the perfect single crystal: either it *cleaves* at the ideal tensile strength after purely elastic straining or it *shears* to produce plastic deformation at the ideal shear strength. Recognizing that cracks will be present, Kelly [2] has examined the ratio (largest tensile stress ÷ largest shear stress) close to the tip of a crack and compared it with the ratio (ideal tensile strength ÷ ideal shear strength). In the absence of mobile dislocations, the relative values of these ratios determine whether cleavage of the crystal occurs ahead of the crack or whether local shear takes place around the crack tip. He concludes that the stress state will favor cleavage if σ_i and τ_i are at all comparable and continues with the interesting observation.[2]

Only in the pure face-centered cubic metals is the ratio of σ_i to τ_i so large that, despite our ignorance of the details and the uncertainty in computing σ_i and τ_i, we can say that theory predicts that cleavage will not occur in the absence of a corrosive environment.

---

[2] In this quotation, Kelley's notation has been changed to the one we are using.

The topic of cleavage mechanisms, and to a lesser extent shear mechanisms of fracture in single crystals and polycrystalline aggregates, is treated in great detail in the metallurgical literature. For our present purpose, which is to describe the microscopic appearance of the fracture surface, we note that we may observe transgranular cleavage, often described due to its appearance as granular or shiny, or a ductile mode of failure which, due to the large local plastic strains, often appears *fibrous* and rough. At higher magnifications ductile fracture in structural metals is seen to be due to the initiation, growth, and coalescence of voids. A pattern of dimples is often observed on the fracture surface. The two modes may appear in combination. For example, in ordinary steels at low temperature, a number of cleaved grains may be joined by regions showing a shear mode of failure. Also, a region where cleavage predominates may contain ligaments which have extended in a ductile manner. A third mode of fracture on the microscale is intergranular separation. This is often observed in long-term failures at elevated temperature and may occur when corrosion abets fracture or when grain boundary embrittlement is present. Here as in a cleavage mode of failure, the overall strain associated with fracture is normally very small. An excellent review of the microscopic modes of fracture with many photographs has been given by Beachem [3].

In the past, the cleavage or brittle mode of fracture on the microscale received greater attention from material scientists than the shear or ductile mode. This is understandable for if a material shows extensive cleavage, it will absorb little energy in fracture and a running crack may propagate catastrophically. Ordinary steels at a low-enough temperature show such behavior, and these materials have provided the motivation for many fracture studies. Another factor is that analytical tools for a study of cleavage mechanisms have been available to material scientists for some time. On the other hand, solutions based on continuum mechanics which permit a detailed study of ductile fracture mechanisms have only recently been developed. The topics of cleavage and ductile failure mechanisms will be treated in Chap. 7.

When cracks are present with Mode I loading and the microscopic mode of fracture is predominantly cleavage, a "square" fracture surface would be expected on a macroscale. By contrast, ductile fracture on a microscale may be associated with either slant or square fracture on a macroscale. For example, in the ordinary tension test, both the "cup" and "cone" parts of the fracture surface may correspond to ductile fracture on a microscale.

At one extreme in the ductile model of failure, certain single crystals slide apart along a single plane until they separate. The strain at final rupture then follows from simple geometric considerations. Similarly, for certain shapes of notched specimens, by neglecting strain hardening and by assuming that sliding occurs along planes of maximum shear stress, expressions may be obtained for the maximum load-carrying capacity and for the extension to *rupture*. Typically, the extension at rupture is of the order of the minimum cross-sectional dimension of the part, although in practice strain hardening will increase this value. These "slip-line" solutions are often useful for predicting the stress and deformation patterns that occur around notches. Some are presented in Chap. 3. Other solutions which are useful in placing a limit on load-carrying capacity are those based on the instability that occurs when strain hardening can no longer balance the effect of decreasing specimen dimensions. It may be shown that if the true stress-true strain curve in tension is described by $\sigma = A\varepsilon^m$ where A and m are constants and elastic strains are neglected, then:

$\varepsilon = m$ at maximum load in tension

$\varepsilon_\theta = m/2$ at maximum pressure-carrying capacity of a thin-walled cylinder with closed ends

$\varepsilon_\theta = m/3$ at maximum pressure carrying capacity of a thin-walled sphere

Here, ε_θ is the hoop strain defined by $\varepsilon_\theta = \ln$ (final diameter $\div$ original diameter). The exponent m may be as low as about 0.05 for quenched and tempered low alloy steels and as high as 0.4 to 0.5 for annealed austenitic stainless steels. More exact expressions can be obtained for elastic-plastic behavior and thick-walled vessels. In these solutions, no fracture considerations are involved, and the material is assumed to have unlimited ductility.

1.3 Historical Aspects

Our subject has its roots in antiquity, for the very survival of our prehistoric ancestors depended on their ability to control fracture to shape stones into tools and weapons. Techniques were developed, during the Stone Age and later, for loading a block of stone and splitting off pieces of a desired shape. The methods may appear primitive, but fracture in a brittle solid is hard to control, and such techniques could only have been developed after a great deal of careful experiment and experience. Much later, in the pre-Columbian Americas, the art of shaping brittle solids became very highly developed. The Incas in Peru matched massive stones with incredible perfection over large areas, while Aztec craftsmen are said to have been able to produce 100 knife blades of obsidian in an hour.

Our recorded experience of fracture begins shortly after steel became available in large quantities in the 1860s. One could argue, somewhat facetiously, that up to that time engineers had a well-developed understanding of fracture control. They had cast iron and masonry which they knew were brittle and wrought iron which was tough. Cast iron bridges, no doubt with generous safety factors, have survived for hundreds of years. Masonry structures, kept everywhere in compression by the skill of a master designer, were used to build Gothic cathedrals. That such construction was possible 400 years before the parallelogram of forces was understood may serve as consolation to the present day designer faced with our as yet inadequate understanding of fracture.

The limitations of wrought iron despite its reputation for toughness were pointed out by Kirkaldy [4] in 1861. He refuted the idea that failures occurred because the metal had "crystallized"—a misconception which persisted for many years. Rather he pointed out that:

the appearance of the same bar may be completely changed from wholly fibrous to wholly crystalline, without calling in the assistance of any of those agents already referred to—viz., vibration, percussion, heat, magnetism, etc.—and that may be done in three different ways;—1st, by altering the shape of the specimen so as to render it more liable to snap; 2nd, by treatment making it harder; and 3rd, by applying the strain so suddenly as to render it more liable to snap from having less time to stretch.

The appendix to this farsighted article contains a number of fascinating editorials and letters to the editor which bemoan the variability in the strength of wrought iron and the inadequacy of the iron used in ship construction.

Returning to steel, the introduction of the Bessemer (1856) and open-hearth (1861) processes led to a steadily increasing use of what had been a rare material. Shortly afterward, occasional reports of catastrophic "brittle" (i.e., low nominal stress) failures began to appear. An interesting summary of some 64 of the early incidents has been given by Shank [5]. He cites a 250 ft. high standpipe 16 ft. in diameter at its base which failed in 1886 during a hydrostatic acceptance test. An observer commented that the failure was due to defective steel plates which were brittle and which unfortunately had been concentrated in the lower part of the tower. No less dramatic was the famous Boston molasses tank which failed in 1919 while containing over two million gallons of molasses. Twelve people were either drowned in molasses or died of injuries, and a portion of the Boston elevated railway was knocked over. After years of testimony the court-appointed auditor ruled that the tank had failed by overstress. The technical testimony must have been long and conflicting. According to Shank, the auditor stated in his decision, "Amid this swirl of polemical scientific waters it is not strange that the auditor has at times felt that the only rock to which he could safely cling was the obvious fact that at least one half of the scientists must be wrong."

The explanations of *faulty material* or *overstress* seem to have been a sufficient description for low-stress brittle failures for most practicing engineers until the period of World War II. Until that time, steels used in general structural engineering were accepted on the results of tensile tests and smooth bar bend tests. However, in aircraft, automobile, and ordnance applications, steels subjected to shock loads were required to pass a notched-bar test. As a result, a certain understanding of notched-bar impact behavior developed in the years from about 1900 onward.

The practice of "nicking" a specimen and striking it with a hammer while held in a vise presumably preceded even the 1862 observations of Kirkaldy. By 1897 a very thoroughly designed impact testing machine was described by Russel [6]. He tested notched specimens in three-point bending and measured energy absorption in the same manner as we do today.

Later, Izod [7] developed an impact test in which a sharply notched specimen was loaded as a cantilever under impact conditions. The International Association of Testing Materials was active in evaluating impact properties. Charpy, who was chairman of the committee on impact testing, proposed another specimen which was loaded in three-point bending and contained a keyhole notch. The specimen he described in 1901 in an unpublished report and later in 1909 [8] had cross-sectional dimensions 30 mm × 30 mm. Later by 1912 the smaller 10 mm × 10 mm cross

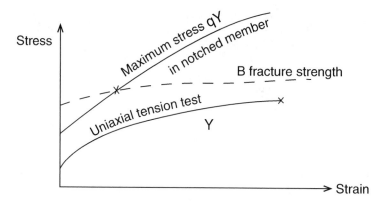

Fig. 1.5 Basis of the Ludwig-Orowan model of notch-brittle behavior

section which we currently use had become accepted. In later years, the practice evolved of testing a Charpy specimen with an Izod-type of V notch so the Charpy-V notch specimen that we use today is a hybrid of these early investigations.

The first attempt to explain the role of a notch in promoting the "ductile-brittle" transition in steels is often attributed[3] to Ludwig and Scheu [9] in 1923. As modified by Orowan [1], it can be explained by reference to Fig. 1.5. In addition to the conventional stress-strain curve Y which culminates in a ductile or fibrous fracture, a steel is assumed to have a brittle strength curve B. Since it lies above Y, only ductile fracture occurs in conventional tension tests. Failure corresponding to curve B is assumed to depend only on the tensile stress. On the other hand, the stress-strain curve Y depends on the shear stresses. Because triaxial tension, such as produced by a notch, decreases the shear stresses, higher loads have to be applied to produce plastic deformation. This effectively raises the curve Y by a factor q and may lead to its intersection with the fracture curve at low strains. The extent to which the flow stress curve can be raised by triaxiality was later quantified by the work of Hencky [15] on plasticity. As pointed out by Orowan, the factor q in Fig. 1.5 can be about 3 in the limit for a tension specimen with a deep circumferential notch. Another solution which is useful in establishing the factor q is that of Bridgman [16] for the necked tension specimen. The early work of Ludwig emphasized the role of triaxial stress in promoting fracture rather than flow. Later, it became realized that low temperatures and high strain rates also elevate the flow stress curve. Thus, the role of the three factors, notches, impact, and temperature, on the ductile-brittle transition in steel is accounted for qualitatively by the Ludwig-Orowan theory. This treatment of fracture is of course quite

---

[3] There is also earlier work which the writer has not yet been able to locate. Orowan refers to the work of Mesnager [11, 12] in 1902 and 1906, which apparently anticipated the ideas of Ludwig and Scheu. An earlier 1909 reference by Ludwig [10] has also been cited by several writers. Modification of Ludwig's concept of notch brittleness was apparently made by Davidenkov [13, 14] in 1936 and 1937 and is discussed by Orowan [1]. The collection of these early references, while not of great practical importance at the present time, would add to our knowledge of the history of fracture.

superficial, and the theory fell into disrepute when dislocation theories of cleavage fracture were developed, starting in about 1948. However, as outlined by Allen [17], these were in turn not completely satisfactory in explaining the fracture transition. Now, with a better understanding of stress fields in cracked and notched specimens, which makes it possible to predict the "brittle" fracture curve, we appear to be returning to the cleavage mode of fracture for explanations not unlike that suggested by the Ludwig-Orowan Mode 1 theory.

Early experience with notched-bar impact tests had shown that the energy required to fracture a steel specimen was very sensitive to temperature. For each steel there was a temperature range in which the energy absorbed would decrease rather suddenly with decreasing temperature. By 1925, over 20 investigations had been carried out on the effect of temperature on iron and steel in the notched-bar impact test. This work was summarized by Greaves and Jones [18].

The convenience of reporting this transition temperature from small standard size specimens seems to have overshadowed other factors. Important observations on the effect of size on fracture behavior were reported for impact tests [19] in 1920. Later in 1935 dramatic differences were reported when geometrically similar Izod specimens were loaded by slow bending [20]. A 10 mm × 10 mm specimen bent 50° without fracturing, while a 100 mm × 100 mm specimen fractured with great violence after bending only 8°. This result was rediscovered in tests on Liberty ship components in World War II and is of obvious importance to anyone wishing to predict structural behavior from laboratory specimens.

On the theoretical side, tools for the analysis of fracture were being developed in the period that we have been discussing, 1900–1940. Unfortunately, some of this work did not find application until much later, and in one case, an important result was completely ignored. This was the work of Wieghardt [21] in 1907. He derived expressions for the elastic stress field around the Mode I crack of Fig. 1.2 and pointed out that the Mode II case could also be treated. This type of solution now forms the basis of what we call linear elastic fracture mechanics. However, at its time, apparently no one saw the use of such a solution for fracture prediction. A few years later, Inglis [22] and Kolosoff [23] obtained solutions for the stresses around elliptical holes. The fact that Inglis published in the proceeding of the Institution of Naval Architects was an omen for the future. At that time, as now, the nominal stress levels in ships, which had been confirmed by measurements on ships at sea, were based on generous safety factors. Despite occasional failures, the Inglis results, which showed that very large stresses could arise at sharp notches under elastic conditions, appeared to attract little attention from those who were designing with steel which was ductile in conventional tests. However, the Inglis solution was utilized by Griffith [24] in what is by now one of the great classic papers in the field of fracture. We will discuss Griffith's work at length in Chap. 4. Basically, he explained the discrepancy between the real and ideal strengths of brittle solids (i.e., elastic to failure) by noting that cracks would be present. Using the Inglis solution and an energy argument, he derived a relation between the stress at fracture and the crack length. An immediate difficulty with the Griffith solution was that the critical crack lengths predicted for most materials were much less than were known to be tolerable. However, by modifying Griffith's approach, while maintaining his basic concept, very useful design procedures have been developed.

A difficulty pointed out by the modified Griffith theory is that the strength-impairing flaws in "brittle solids" are very small. Generally, they cannot be detected by nondestructive testing. We are then faced with the dilemma that nominally identical specimens of, for example, a ceramic will contain different-sized flaws and will fail at different stress levels. This problem was treated by Weibull [25], with great physical insight, in a series of papers in 1939. He developed a probabilistic treatment of the strength of brittle solids under tensile states of stress. Again, as with Griffith, Weibull's work seems to have been ahead of its time and slow in gaining recognition. It is now used extensively in work with ceramics and in fatigue and reliability studies. Some of its implications will be examined in Chap. 8.

The events we have outlined have led us to the beginning of World War II or what might the termed, with a narrower viewpoint, the era of welded construction. Just prior to World War II, about 50 bridges were built across the Albert Canal in Belgium. Quoting Shank [5], three of these failed in a 2-year period from 1938–1940. As might be expected, the local steel industry claimed that the steel was above reproach and the quality of the welds was at fault. The welding representatives, on the other hand, were satisfied that the failure was not due to weakness of imperfections in the welded joints. Later, a German investigation showed that most failures initiated at welds and that many welds were defective. Notch impact tests on material from the failures showed that most specimens were at least partially brittle at the failure temperature. Their report, which seems to have been thorough, attributed failures to (a) multiaxial restraint and residual stress, (b) low ambient temperature, and (c) the low notch impact characteristics of the steel.

At the same time, the construction of thousands of welded ships led to an appalling number of catastrophic failures. A study of about 4000 welded ships shows 200 listed as Group I failures. These are classified as those that sank or were left in a dangerous condition due to fracture. A very complete summary of the welded ship problem, as well as an account of the earlier failures, has been given by Parker [26]. Attempts to reproduce these ship failures in the laboratory led to a great deal of difficulty. However, a correlation was found that at the failure temperature, the plates in which failure started had significantly lower energy absorption in notched-bar tests than plates in which failure had not occurred. Although the ship fracture problem was still not understood, this finding gave emphasis to the transition temperature in notched-bar impact as an index of steel quality and as a design rule. As a result a great deal of valuable research was done on metallurgical techniques for lowering the transition temperature, and many different types of tests were developed which defined and measured a transition temperature.

We will turn again to this topic in Chap. 6. However, since the description of the ductile-brittle transition in steel by the Charpy-V notch test has become so widely accepted by engineers, it may be well to dwell on some of its values and limitations. An advantage of this approach is that it enables an enormous amount of data to be collected on different steels using an inexpensive test, and the effect of various variables—composition, grain size, etc.—can be evaluated by their effect on this single quantity, the transition temperature. However, this temperature can be defined in a variety of ways using energy absorption, fracture appearance, or contraction at the notch root, and the choice of the definition as well as the type of test may change the relative rating of different steels.

A limitation of the transition temperature approach in design is that in measuring this quantity with laboratory specimens, it is possible to get brittle behavior only by combining sharp notches and impact loading with low temperature. These severe conditions do not necessarily simulate those existing in the real structure which explains the fact that numerous structures have operated satisfactorily for many years at temperatures well below their transition temperature as measured by standard tests. Also, there is the dilemma that when brittle fracture of ordinary steel in laboratory tests first appears at the transition temperature, the nominal stresses are at least equal to the yield strength. By contrast, catastrophic fracture of welded ships and other steel structures has occurred at nominal stresses well below yield values. A major contribution to understanding this difference in behavior was provided by Robertson's [27] studies of the conditions for arrest of a running crack. He showed that a crack once initiated could propagate at stress levels very much lower than required for crack initiation. Thus, the problem becomes one of explaining crack initiation at low nominal stresses rather than its subsequent catastrophic propagation through a structure. Although low-stress initiation in ordinary low-carbon steel is still not fully understood, two factors can be identified as major contributors. One is the effect of size; the other is welded construction. In geometrically similar notched parts, we will see that one would not expect a size effect if the microscopic fracture mode is cleavage. However, if failure initiates by ductile fracture on a microscale, which may be followed by cleavage as the crack starts to propagate, then it is possible for small specimens and large parts to show the dramatic difference we discussed earlier in connection with slow-bend tests on Izod-type specimens.

Turning to welded construction, this may produce large residual stresses and local changes in fracture properties. These two variables are not easily separated, but, together or individually, they may lead to fracture of non-stress-relieved welded steel structures at low nominal stresses, as we will describe in Chap. 6.

Although the work outlined pointed out inadequacies in the transition temperature approach, the real incentive for a more rational design approach was provided by dramatic failures in 3 different areas in the 1950s. The first of these was the catastrophic failure while climbing to cruising altitude of two de Havilland Comet airliners in January and April of 1954. The exhaustive investigation [28] that followed traced the failures to cracks in the pressurized fuselage. These initiated at stress concentrations and grew during pressurization cycles until they became long enough to propagate catastrophically. In retrospect, the report [25] is somewhat lacking in that it ignored the final, catastrophic, fast fracture phase of the failure. However, the fact that a structure made of a ductile aluminum alloy, which shows no transition temperature, could undergo catastrophic crack propagation at nominal stress levels well below yield was an important lesson from the Comet failures. In the same period from January 1953 to March 1956, an alarming series of some six failures occurred in steam turbo-generator rotors. Again an extensive investigation was undertaken, and the results were reported in 1958 [29, 30]. This and subsequent work in the same area have been summarized by Yukawa, Timo, and Rubio [31]. Once more, the effect of size on fracture behavior was noted, and spin tests to simulate service failures were correlated with other tests. Attention was given to the slow stable growth of defects by low-cycle fatigue, as well as the presence of hydrogen. This emphasized the need for both better nondestructive testing techniques and better control of steel quality in forgings. It appears that these studies marked the first major industrial application of an analytical approach to fracture control. They stimulated a great deal of subsequent interest, as well as confidence, in this type of calculation.

The third problem area in the 1950s arose with the development of high-strength steel casing for rocket motors. Many catastrophic failures occurred in proof-testing at nominal stress levels below yield, but again these materials showed no sudden change of energy absorption with temperature in impact tests. The severity of this problem led to the formation of an ASTM committee in 1959 to study test methods for fracture control. The subsequent work of this committee, which will be outlined in Chap. 5, provides an interesting and well-documented viewpoint on the development of test methods during the succeeding years.

Fortunately, on the analytical side, the modification of Griffith's original energy approach to fracture prediction by Irwin leads to very useful quantitative methods for relating the stress at failure to flaw size in situations where fracture precedes general yielding. This approach which was proposed by Irwin [32] in 1947 and later expanded upon by Irwin and Kies [33] in 1952 at first was met with opposition but is now firmly established. It forms the basis of our present linear elastic fracture mechanics (LEFM). While Irwin and others were developing the energy approach to fracture prediction, workers [34] in the Soviet Union in the 1950s were predicting fracture in geological applications by using the "stress intensity factors" which describes essentially the magnitude of the state of stress near the tip of a crack for elastic behavior. In Fig. 1.6 the stress along the extension of the crack in the direction of loading ($\sigma_y(x)$ for $y = 0$) is expressed in terms of the stress intensity factor K which can be computed for a wide variety of cracked configurations. The path surrounding the part and labeled G indicates that the original Griffith-Irwin approach to fracture prediction required a calculation, involving the entire part, of the energy G released by unit increase in crack area. These two analytical approaches to fracture prediction were unified by Irwin [35] in a series of papers from 1955 to 1957 where he pointed out that G and K were related by a simple expression for through cracks in plane specimens. This was a great step forward for it combined the intuitive appeal and past experience of the energy approach with the mathematically simpler problem of determining stress intensity factors. As a result in the following years and up to the present day, a great deal of fracture research has centered on the determination of stress intensity factors by analytical, numerical, and experimental methods.

In one sense, in retrospect, the emphasis on stress intensity factors was unfortunate in that this concept is restricted to elastic behavior, while the energy approach can be applied, at least in principle, for more general material behavior. Interest in applying

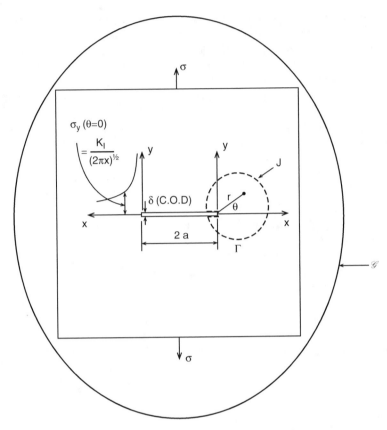

Fig. 1.6 A cracked plate showing schematically four approaches to fracture prediction. These are the global energy balance G, the energy approach applied to a control volume at the crack tip J, the stress intensity factor K, and the crack opening displacement δ

the energy approach to situations involving large-scale or general yielding was stimulated by Rice's [36] observation in 1968 that the rate at which energy is released by crack growth could be obtained from a path-independent integral taken around the crack tip, as sketched in Fig. 1.6. This quantity which he termed J should apply for non-linear elasticity and, hence, by analogy to deformation theory plasticity. For linear elastic behavior, J is identical to G and in fact can be shown to be the consequence of applying the original Griffith concept to a control volume at the crack tip rather than to the entire specimen. Very many experimental studies have been devoted to the measurement of J, but unfortunately much less attention has been given to its physical significance when large-scale or general yielding is involved. Curiously, although Rice's path-independent integral motivated this work, it is not employed in the usual methods for obtaining J experimentally. Starting in 1972 [37] various methods have been developed based on the experimental determination of non-linear load-deflection curves for cracked specimens. The same approach was used in 1953 for non-linear rubbery polymers [38]. The almost 20-year lag in applying it to metals indicates a rather stratified viewpoint in research.

An earlier attempt to develop what is often called "general yield fracture mechanics" (GYFM) has grown from the suggestion by Wells [39] in 1951 that since the fracture process depends on the large strains in a small zone ahead of the crack, a displacement criterion should be involved. The "crack opening displacement" (δ or COD) shown in Fig. 1.6 can be related to G and K for linear elastic behavior but can, in principle, be measured for any type of material behavior. This concept has been studied extensively, particularly in Great Britain. Close correspondence has been shown by Wells [40] and others between a suitably defined crack opening displacement and the value of J computed from Rice's path-independent integral. The use and limitations of the J and COD approaches will be discussed in Chap. 6.

In attempting a concise historical review of some of the main developments in fracture, we have had to be qualitative and have omitted many contributions such as the application of LEFM to correlate fatigue crack growth studies and recent developments in the analysis of ductile fracture. These and many other facets of fracture will be dealt with in Chaps. 4 through 8 in more quantitative terms. The next two chapters provide some background on the equations of solid mechanics and list some analytical solutions which are useful in studying fracture.

References

1. Orowan, E. Notch Brittleness and the Strength of Metals, Trans. Inst. of Engrs. Shipbuilders Scotland, Vol. 89, 1945–1956, p. 165–215. See also, Fracture and Strength of Solids, "Reports in Progress in Physics," Vol. XII, 1949, p. 185–232
2. Kelly A. Strong solids. Oxford: The Clarendon Press; 1966.
3. Beachem CD. Microscopic fracture processes. In: Liebowitz H, editor. Fracture, vol. 1. New York: Academic; 1968. p. 243–349.
4. Kirkaldy D. Results of an experimental inquiry into the comparative tensile strength and other properties of various kinds of wrought-iron and steel. Proc. Scot. Shipbuilders Assoc. (Glasgow) 1860–1861, p. 1–187. With Appendix pp. 189–212.
5. Shank ME. A critical survey of brittle failure in carbon plate steel structures other than ships, Welding Res. Council Bull., No. 17, Jan. 1954. A condensation is given in: Brittle Failure of Nonship Steel-Plate Structures. Mech Eng. 1954;76:23–8.
6. Russel SB. Experience with a new machine for testing materials by impact. Proc Am Soc Civil Engrs. 1897;23:550–7.
7. Izod EG. Testing brittleness of steel. Engineering. 1903;76:431–2.
8. Charpy G. Report on Impact Tests of Metals, Proc. Int. Assoc. for Testing Materials, Vol. 1, Report III, (1909): Report on Impact Tests and the Work of Committee No. 26, Proc. 6th Congress Int. Assoc. for Testing Materials, Paper IV (1912).
9. Ludwig P, Scheu R. Ueber Kerbwirkungen bei Flusseisen. Stahl und Eisen. 1923;43:999–1001.
10. Ludwig P. Elemente der Technologischen Mechanik. Berlin: Springer; 1909.
11. Mesnager A. Reunion de Membres Francais et Belges de l'Association Internaltionale des Methode d'Essais, Dec. 1902, p, 395–405. Quoted by E. Orowan in "Design of Piping Systems" Chapter 1, published for the M.W. Kellog Company by John Wiley and Sons 1956.

12. Mesnager, A. Internat. Assoc. for Testing Materials, Brussels Congress 1906, Rapport non-offic. No. A6f, pp. 1–16. Quoted by D.K. Felbeck and E. Orowan in Experiments on Brittle Fracture of Steel Plates, Welding Journal, 34, 1955, p. 570s–575s

13. Davidenkov NN. Dinamicheskaya Ispytania Metallov. Moscow 1936, Quoted by E. Orowan in Ref. 1.

14. Davidenkov NN, Wittman F. Mechanical analysis of impact brittleness. Phys Tech Inst (USSR). 1937;4:308. Quoted by E. Orowan in the same source as for Ref. 10a and by N.P. Allen, Ref. 14, as Report FE4/93, Phys. Tech. Inst. Leningrad 1937.

15. Hencky H. Uber einige statisch bestimmte Falle des Gleichgewichts in plastischen Korpern. ZF Angewandte Math und Mech. 1923;3:241–51.

16. Bridgman PW. The stress distribution at the neck of a tensile specimen. Trans Am Soc Metals. 1944;32:553–74. See also his book: "Large Plastic Flow and Fracture", McGraw-Hill Book Co., New York 1952.

17. Allen NP. Studies of the brittle fracture of steel during the past twenty years. Int J Fracture Mech. 1965;1:245–71.

18. Greaves RH, Jones JA. The effect of temperature on the behaviour of iron and steel in the notched-Bar impact test. J Iron and Steel Inst. 1925;112:123–65.

19. Stanton TE, Batson RGC. On the Characteristics of Notched-Bar Impact Tests, Minutes of Proc. of Inst. of Civil Engrs. (London) 211, 1920–1921, Pt. 1, pp. 67-100, and figure facing p. 408.

20. Docherty JG. Slow bending tests on large notched bars. Engineering. 1935;139:211–3.

21. Weighardt K. Uber das Spalten und Zerreissein elasticher Korper. Z F Math U Physik. 1907;55:60–103.

22. Inglis CE. Stresses in a plate due to the presence of cracks and sharp corners. Trans Inst of Naval Architects (London). 1913;55:219–41.

23. Kolossoff, G. Uber einige Eigenschaften des ebenen Problems der Elastizitatstheorie, Z. F. Math U. Physik, 62, 1913–1914, p. 284–409.

24. Griffith AA. Phenomena of flow and rupture in solids. Phil Trans Roy Soc (London). 1920;221A:163–98.

25. Weibull W. A Statistical theory of the strength of materials (No. 151) 45 pages and "The Phenomenon of Rupture in Solids" (No. 153), 55 pages, Ingeniorsvetenskaps academiens, Stockholm, Sweden, 1939.

26. Parker ER. Brittle behavior of engineering structures. New York: John Wiley and Sons; 1957.

27. Robertson TS. Propagation of brittle fracture in steel. J Iron and Steel Inst. 1953;175:361–74.

28. "Civil Aircraft Accident"—Report of the Board of Inquiry into the Accidents to Comet G-ALYP on 10th January 1954, and Comet G-ALYY on 8th April 1954, London HMSO 1955.

29. Winnie DH, Wundt BM. Application of the Griffith-Irwin theory of crack propagation to the bursting behavior of disks, including analytical and experimental studies. Trans Am Soc Mech Engrs. 1958;80:1643–58.

30. Lubahn JD, Yukawa S. Size effects in slow notch-bend tests of a nickel-molybdenum-vanadium steel. Proc ASTM. 1958;58:661–77.

31. Yukawa S, Timo DP, Rubio AD. Fracture design practices for rotating equipment. In: Liebowitz H, editor. Fracture, vol. 5. New York: Academic Press; 1969. p. 65–157.

32. Irwin GR. Fracture dynamics, p. 147–166. In: "Fracturing of Metals" Am. Soc. for Metals, Cleveland 1948.

33. Irwin GR, Kies JA. Fracturing and fracture dynamics. Welding J. 1952;31:95s–100.

34. Barenblatt GI. The mathematical theory of equilibrium cracks in brittle fracture. Adv Appl Mech. 1962;7:55–129. pub. by Academic Pres. (see this article for references to earlier work in Russian dating back in 1955).

35. Irwin GR. Analysis of stresses and strains near the end of a crack traversing a plate. J Appl Mech. 1957;24:361–4. See also, Irwin, G.R., Onset of Fast Crack Propagation in High Strength Steel and Aluminum Alloys, Proc. Sagamore Conference on Strength Limitations of Metals, Part 2, 1955, p. 289–305.

36. Rice JR. A path independent integral and the approximate analysis of strain concentration by notches and cracks. J Appl Mech. 1968;35:379–86.

37. Begley JA, Landes JD. The J integral as a failure criterion. Philadelphia: ASTM; 1972. p. 1–23. in ASTM Spec. Tech Pub. 514.

38. Rivlin RS, Thomas AG. Rupture of rubber. I. Characteristic energy of tearing. J Polymer Sci. 1953;10:291–318.

39. Wells AA. Unstable crack propagation in metals: cleavage and fast fracture. Proc Crack Propagation Symposium, Cranfield, England. 1961;1:210–30. pub. by The College of Aeronautics, Cranfield, England, 1962.

40. Wells AA. Fracture control: past, present and future. Exp Mech. 1973;13:401–10.

Chapter 2
Stress, Strain, and the Basic Equations of Solid Mechanics

2.1 Introduction

In this chapter our objective will be to review the concepts of stress and strain and to present the equations which relate these quantities, and if necessary other variables, for various types of material behavior. Some general comments will be made on the methods of solution of the equations, but a description of specific solutions useful for fracture prediction will be deferred to the next chapter. For this reason, the reader with background in the analysis of stress and strain and the equations for elastic and plastic deformation can proceed to Chap. 3. The treatment in this chapter is based on a "homogeneous" and "isotropic" continuum. That is, the properties do not vary with location and are the same in all directions. Although individual grains in a ceramic or metal may not be isotropic, the assumption of isotropic behavior is justified for a randomly oriented polycrystalline aggregate if the dimensions over which stress and strain change appreciably are largely compared to those of the individual grains. A similar argument applies to noncrystalline materials if grain size is replaced by the dimension over which significant microstructural changes occur. However, for small regions the assumption of a homogeneous isotropic continuum is less realistic. In subsequent chapters, solutions based on this concept will be applied to predict stresses at the tips of sharp cracks and the growth of small voids. In both cases the dimensions involved may be smaller than the grain size. This approach is justified by expediency, since more realistic calculations are vastly more complicated, but its limitations should not be overlooked.

Whatever the type of material behavior being considered, we deal with a continuous, deformable body subject to combinations of surface and body forces which may be static or may vary with time. These forces are transmitted through the body and it deforms in some manner.

To describe the way that the forces are distributed, we use the concept of stress at a point. The stresses are related to the surface and body forces by consideration of equilibrium or through Newton's law if the loading is dynamic. The resulting

© Springer Science+Business Media New York 2016
C.K.H. Dharan et al., *Finnie's Notes on Fracture Mechanics*,
DOI 10.1007/978-1-4939-2477-6_2

deformations in the body are described using the concept of strain at a point. The strains must be compatible so that continuity is satisfied and are related to the displacements by geometry. By themselves, stress and strain have nothing to do with material behavior. Thus, it is meaningless to refer to "elastic stress" or "plastic stress." However, often it is convenient to separate strain into elastic and plastic components, but the relationship of the total strain to the displacements is purely geometric.

2.2   Analysis of Stress

We consider the load δF acting across a small plane area δA in a solid body. By taking the components of the load normal and tangential to the plane and letting δA tend to zero, we find the normal and tangential stresses at a point referred to a plane and it has a direction relative to that plane.

- Many different systems of notation have been used to describe the stresses. We will use the one shown in Fig. 2.1 in which σ_x, σ_y, σ_z are the normal stresses on the faces of a small cube oriented in the x, y, and z directions. The shear stresses on the faces are denoted by τ_{xy}, τ_{yz}, τ_{zx}. One can adopt the convention that the first subscript indicates the normal to the plane on which the shear stress acts, while the second subscript indicates the direction of the shear stress. However, moment equilibrium[1] requires $\tau_{xy} = \tau_{yx}$; $\tau_{yz} = \tau_{zy}$; $\tau_{zx} = \tau_{xz}$; so the subscript convention for shear stresses is not important for our purposes. Normal stresses are taken as positive when tensile and the positive direction of shear stresses is shown in the figure. Although the Cartesian coordinate system of Fig. 2.1 is the most commonly used, it is often more convenient to describe a state stress with respect to three orthogonal directions which are more appropriate to the shape being considered. Later, we will discuss polar, spherical, and elliptical coordinates as they are needed. The equations for transformation of stress and strain from one coordinate system to another are developed in many texts and will not be discussed in this introductory treatment.

The majority of practical problems involve a two dimensional state of stress in which $\tau_{xy} = \tau_{yz} = 0$ and σ_z is a principal stress which may or may not be zero. The remaining stress components are σ_x, σ_y, and τ_{xy} as shown in Fig. 2.2. By examining all possible orientations of a small element in the x–y plane, one is found for which the shear stresses vanish, and the normal stresses have the maximum and minimum values for all possible orientations of the element. These normal stresses are referred to as the principal stresses σ_1, σ_2. Equations relating to σ_1, σ_2, and the

---

[1] This ignores the possible presence of distributed moments on the faces of the unit cube. Although a large amount of literature has appeared on this topic, the distributed moment does not appear to be important in practice.

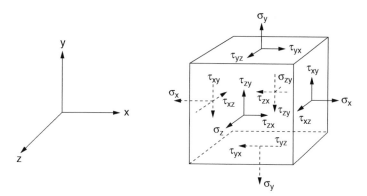

Fig. 2.1 Notation used for stresses

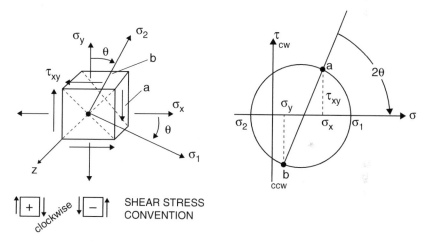

Fig. 2.2 Mohr construction

angle θ shown in Fig. 2.2 are derived in texts on Strength of Materials. A useful graphical interpretation of these equations is the Mohr construction in which the normal and shear stresses on a plane are plotted as a point on a circle. In drawing the Mohr circle, sign conventions are needed and arbitrarily tensile stresses are taken as positive and "clockwise" shear stresses as positive. We should note that the sign convention for shear stresses is only for the Mohr circle and that later in Chap. 8 in discussing the Mohr-Coulomb theory of failure under compressive state of stress, it will be more convenient to take compressive stresses as positive. The essential aspects of the Mohr construction, once the sign convention is selected, is that points on the Mohr circle represent the states of stress on planes whose normals lie in the x–y plane and that the angle between two points on the circle is twice the angle between the actual planes.

From Fig. 2.2, the principal stresses σ_1, σ_2 are seen to be

$$\sigma_1, \sigma_2 = \frac{\sigma_x + \sigma_y}{2} \pm \left\{ \left(\frac{\sigma_x - \sigma_y}{2} \right)^2 + \tau_{xy}^2 \right\}^{1/2} \tag{2.1a}$$

The angle θ between σ_x and σ_1 is given by

$$\tan 2\theta = \tau_{xy} \div \frac{\sigma_x - \sigma_y}{2} \tag{2.1b}$$

For all possible orientations of the element, the maximum value of the shear stress on its faces is the radius of the Mohr circle:

$$\tau = \frac{\sigma_x - \sigma_y}{2} = \left\{ \left(\frac{\sigma_x - \sigma_y}{2} \right)^2 + \tau_{xy}^2 \right\}^{1/2} \tag{2.1c}$$

This may not be the maximum shear stress in the material for if σ_1 and σ_2 have the same sign with $\sigma_1 > \sigma_2$ and $\sigma_x = \sigma_3$ is zero or has opposite sign to σ_1 and σ_2, then

$$\tau_{max} = \frac{\sigma_1 - \sigma_3}{2}$$

As an illustration, consider the thin-walled cylindrical vessel with closed ends carrying internal pressure shown in Fig. 2.3. The term "thin-walled" means that the inside radius R is at least an order of magnitude greater than the wall thickness h. Thus, the variation of stress across the wall may be neglected, and the stresses can be found from consideration of static equilibrium. It is natural here to replace the Cartesian coordinate system by one based on the circumferential, θ; axial, a; and radial, r, directions. By the symmetry of the loading, σ_θ, σ_a, σ_r are the principal stresses, and the shear stress components $\tau_{\theta a}$, τ_{ar}, $\tau_{r\theta}$ vanish. By considering static equilibrium, it may be shown that

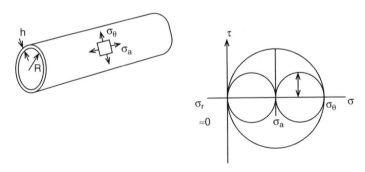

Fig. 2.3 Mohr circles for thin-walled cylinder under internal pressure

$$\sigma_\theta = \frac{pR}{h} \quad \text{and} \quad \sigma_a = \frac{pR}{2h}$$

Since $R \gg h$ and σ_r varies from $-p$ at the inside wall to zero at the outside wall, it is usually assumed that $\sigma_r \simeq 0$. These stresses are said to be statically determinate since they may be calculated from statics without any reference to material behavior. By contrast, the stresses in a thick-walled cylinder are statically *indeterminate* and depend on material behavior, being quite different, for example, for elastic and plastic behavior. Returning to the thin-walled cylinder, we can draw three Mohr circles and find the three principal shear stresses $(\sigma_\theta - \sigma_a)/2$, $(\sigma_a - \sigma_r)/2$, $(\sigma_r - \sigma_\theta)/2$. It is seen that if we considered only the σ_a, σ_θ, the maximum shear stress in the material would be underestimated by a factor of 2.

If the problem is not two dimensional, then an analytical solution is more convenient than graphical methods. Given $\sigma_x, \sigma_y, \sigma_z, \tau_{xy}, \tau_{yz}, \tau_{zx}$, these can be used directly to predict elastic and plastic deformation or the onset of yielding without determining the principal stresses $\sigma_1, \sigma_2, \sigma_3$. However, it is sometimes useful to determine the principal stresses, and on rare occasions it is necessary to calculate their directions relative to the x, y, and z axes. We recall that the principal stresses act on planes upon which the shear stresses are zero. One principal stress is the greatest, and another is the least normal stress for all possible orientations of the elemental cube at a given point in the material. Considering the tetrahedral element shown in Fig. 2.4, it is assumed that a principal stress S acts on the plane ABC. The normal n to this plane makes angles with the x, y, and z axes, and the cosines of these angles are denoted by n_x, n_y, n_z. The areas of the planes shown in Fig. 2.4 are related to the area of plane ABC through these direction cosines, e.g., $AOC = ABC\ n_x$. Equilibrium of forces in the x direction requires

$$\sigma_x AOC + \tau_{yx} AOB + \tau_{zx} COB = S n_x ABC$$

Similar equations may be written for the y and z directions. By expressing the areas in terms of area ABC and the direction cosines, the three equations become

$$
\begin{aligned}
(\sigma_x - S)n_x + \tau_{yx}n_y + \tau_{xz}n_z &= 0 \\
\tau_{xy}n_x + (\sigma_y - S)n_y + \tau_{zy}n_z &= 0 \\
\tau_{xz}n_x + \tau_{yz}n_y + (\sigma_z - S)n_z &= 0
\end{aligned}
\tag{2.2a}
$$

The subscript convention for the directions of the shear stresses is retained in Eq. (2.2a) to emphasize the directions of the forces on the tetrahedral element in Fig. 2.4, but we recall that $\tau_{xy} = \tau_{yx}$, etc.

These three equations relating the three direction cosines n_x, n_y, n_z have a solution only if the determinant of the coefficients vanishes. That is,

Fig. 2.4 Element used to determine principal stresses

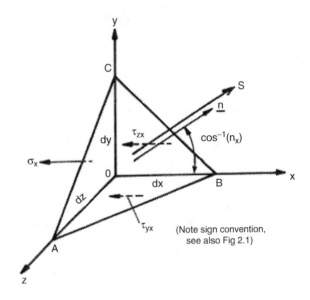

$$\begin{vmatrix} (\sigma_x - S) & \tau_{yx} & \tau_{xz} \\ \tau_{xy} & (\sigma_y - S) & \tau_{zy} \\ \tau_{xz} & \tau_{yz} & (\sigma_z - S) \end{vmatrix} = 0 \qquad (2.2b)$$

The three values of S satisfying this cubic equation are the three principal stresses $\sigma_1, \sigma_2, \sigma_3$. Expanding the determinant

$$S^3 - I_1 S^2 + I_2 S - I_3 = 0$$

$$\text{where} \quad I_1 = \sigma_x + \sigma_y + \sigma_z,$$
$$I_2 = -\left(\sigma_x \sigma_y + \sigma_y \sigma_z + \sigma_z \sigma_x\right) + \tau_{xy}^2 + \tau_{yz}^2 + \tau_{zx}^2$$
$$I_3 = \begin{vmatrix} \sigma_x & \tau_{xy} & \tau_{xz} \\ \tau_{xy} & \sigma_y & \tau_{yz} \\ \tau_{xz} & \tau_{yz} & \sigma_z \end{vmatrix}$$

The three quantities I_1, I_2, I_3, often referred to as "invariants," cannot depend on the choice of the x, y, and z axes since an infinite variety of these directions correspond to given magnitudes and directions of the principal stresses. Thus, $I_1 = \sigma_1 + \sigma_2 + \sigma_3$, $I_2 = -(\sigma_1 \sigma_2 + \sigma_2 \sigma_3 + \sigma_3 \sigma_1)$, and $I_3 = \sigma_1 \sigma_2 \sigma_3$. The directions of the principal stresses relative to the x, y, and z axes may be found from Eq. (2.2a) once the principal stresses are known. Actually, only two of these three equations are needed since the direction cosines satisfy the equation $n_x^2 + n_y^2 + n_z^2 = 1$.

In discussing plastic deformation and fracture, we often refer to the average normal stress

$$p = \frac{I_1}{3} = \frac{\sigma_x + \sigma_y + \sigma_z}{3} = \frac{\sigma_1 + \sigma_2 + \sigma_3}{3} \tag{2.3}$$

as the "hydrostatic stress" even though its value may be positive (i.e., tensile). We should not confuse this quantity with a purely hydrostatic state of stress in which $\sigma_1 = \sigma_2 = \sigma_3$ (i.e., equal triaxial tension or compression). Since plastic deformation occurs by shear, a state of stress $\sigma_1 = \sigma_2 = \sigma_3$ will not produce yield or plastic deformation. By the same token, the addition or subtraction of a given stress to all normal stresses will not influence yield or plastic deformation. However, this may have a profound effect on fracture. It is convenient for our subsequent discussion of the yield criterion to subtract the quantity p (perhaps better referred to as the "hydrostatic component of stress") from all normal stresses to obtain what are called the stress deviators:

$$\sigma'_x = \sigma_x - p; \quad \sigma'_1 = \sigma_1 - p; \quad \text{shear stresses unchanged}$$

The principal stress deviators $\sigma'_1, \sigma'_2, \sigma'_3$ are obtained from a determinant similar to that for the principal stresses. Thus, $\sigma'_1, \sigma'_2, \sigma'_3$ are obtained from

$$S'^3 - J_1 S'^2 + J_2 S' - J_3 = 0$$

where $J_1 = \sigma'_x + \sigma'_y + \sigma'_z = 0,$

$$J_2 = \left(\sigma'_x \sigma'_y + \sigma'_y \sigma'_z + \sigma'_z \sigma'_x \right) + \tau^2_{xy} + \tau^2_{yz} + \tau^2_{zx}$$

$$J_3 = \begin{vmatrix} \sigma'_x & \tau_{xy} & \tau_{xz} \\ \tau_{xy} & \sigma'_y & \tau_{yz} \\ \tau_{xz} & \tau_{yz} & \sigma'_z \end{vmatrix}$$

The stresses must balance any applied surface stresses on the boundary of the body. Considering the plane ABC of Fig. 2.4 as an element to the boundary with surface forces per unit area P_x, P_y, P_z in the x, y, and z directions, we have

$$\sigma_x n_x + \tau_{xy} n_y + \tau_{xz} n_z = P_x \tag{2.4a}$$

Equilibrium must also be satisfied internally. Looking at the element in Fig. 2.5 and considering balance of forces in the x direction,

$$\frac{\partial \sigma_x}{\partial x} + \frac{\partial \tau_{xy}}{\partial y} + \frac{\partial \tau_{xz}}{\partial z} = 0 \tag{2.4b}$$

When body forces F_x, F_y, F_z per unit volume are present and if the displacements u, v, and w in the x, y, and z directions correspond to significant accelerations $\partial^2 u / \partial t^2$, etc., then the equilibrium equations become

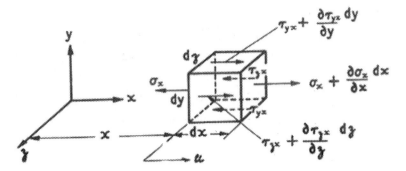

Fig. 2.5 Element showing components in x direction

Fig. 2.6 Element used to derive equation of radial equilibrium

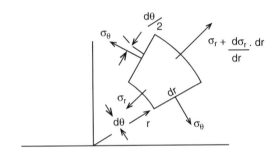

$$\frac{\partial \sigma_x}{\partial x} + \frac{\partial \tau_{xy}}{\partial y} + \frac{\partial \tau_{xz}}{\partial z} + F_x = \rho \frac{\partial^2 u}{\partial t^2} \tag{2.4c}$$

where t is time and ρ is the mass density. Similar equations of (2.4a), (2.4b), and (2.4c) may be written for the y and z directions.

Earlier a polar coordinate system was used in discussing a thin-walled pressure vessel. If $\sigma_\theta, \sigma_r, \sigma_a$ are the principal stresses, which means that $\tau_{r\theta} = \tau_{\theta z} = \tau_{zr} = 0$, then the radial equilibrium of stresses in the absence of body forces or radial acceleration may be obtained from Fig. 2.6 as

$$\left(\sigma_r + \frac{\partial \sigma_r}{\partial r} dr \right)(r + dr)d\theta - \sigma_r r d\theta - 2\sigma_\theta dr \frac{d\theta}{2} = 0 \quad \text{or}$$

$$r\frac{d\sigma_r}{dr} + \sigma_r - \sigma_\theta = 0 \tag{2.5}$$

2.3 Analysis of Strain

Strain is a purely geometrical description of the change in size and shape of an element in a deforming body, relative to some reference state. For elastic behavior in all but rubbery polymers or microscopic whiskers, the strains are small and we may use the concept of "infinitesimal strain." The analysis of strains is involved and, in general, becomes rather complicated, but the special case of "homogeneous strain" is relatively straightforward. In this case straight lines before straining remain straight and parallel lines remain parallel although their direction may be altered. A concise discussion of homogeneous strain has been given by Jaeger [1]. Here we will consider only the simple cases of uniaxial extension or compression, the deformation of a rectangular block into another rectangular block, the axisymmetric expansion of a cylinder, and simple shear.

The uniaxial extension of a rod from length L_0 to length L illustrates that there is no one unique definition of strain, but rather a definition is chosen which is convenient for the problem being solved. The most familiar definition is the "engineering strain," $\varepsilon = (L - L_0)/L_0$, the fractional change in the initial length. In work with finite strains, a definition often used is $\varepsilon = (L/L_0)^2$, while in calculations involving plastic deformation, it is convenient to use the "logarithmic" strain $\varepsilon = \ln(L/L_0)$, also known as the "true" or "natural" strain. For small strains the engineering strain and logarithmic strain are almost identical. For large strains such as may arise in metal working, the experimental result that essentially no volume change occurs in plastic deformation makes the logarithmic strain definition a convenient one. To illustrate this, if the deformation shown in Fig. 2.7 occurs without change in volume (i.e., if elastic strains are neglected),

$$X_0 Y_0 Z_0 = XYZ \quad \text{or} \quad \frac{X}{X_0} \frac{Y}{Y_0} \frac{Z}{Z_0} = 1$$

The engineering and logarithmic strain definitions may be applied in the x, y, and z directions to give $X/X_0 = (1 + \varepsilon_z)$ for engineering strain and $X/X_0 = e^{\varepsilon_z}$ for logarithmic strain with similar expressions in the y and z directions. The constancy of volume condition then becomes

$$(1 + \varepsilon_x)(1 + \varepsilon_y)(1 + \varepsilon_z) = 1 \quad \text{Eng. Strain}$$
$$\varepsilon_x + \varepsilon_y + \varepsilon_z = 0 \quad \text{Log. Strain}$$

Thus, by adopting the logarithmic strain definition, if only plastic deformation is being considered, one normal strain may be expressed in terms of the other two.

Another advantage of the logarithmic strain definition arises if interrupted deformation is being considered. As shown in Fig. 2.8, if deformation is carried out from L_0 to L, the engineering strain definition requires the original length L_0 to

Fig. 2.7 Constant volume deformation

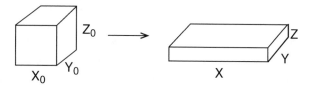

Fig. 2.8 Two definitions of strain

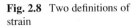

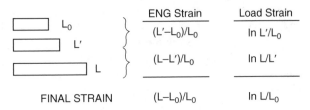

	ENG Strain	Load Strain
	$(L'-L_0)/L_0$	$\ln L'/L_0$
	$(L-L')/L_0$	$\ln L/L'$
FINAL STRAIN	$(L-L_0)/L_0$	$\ln L/L_0$

Fig. 2.9 Coordinates used for relating strains to displacements in axisymmetric deformation

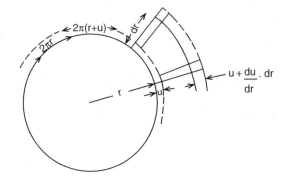

be used in the second step, while with logarithmic strain the current length L' is always taken as the "gauge length." It is also found that the stress–strain curves for annealed ductile metals in tension and compression coincide only if the "true stress" (load divided by current cross-sectional area) is plotted as a function of the true (logarithmic) strain.

Another special case is the axisymmetric deformation shown in Fig. 2.9 in which the radial and circumferential strains may be expressed in terms of the radial displacement u. Using the engineering strain definition,

$$\varepsilon_\theta = \frac{2\pi(r+u) - 2\pi r}{2\pi r} = u/r; \quad \varepsilon_r = \left(\frac{du}{dr}dr\right)/dr = \frac{du}{dr}$$

For the logarithmic strain definition,

$$\varepsilon_\theta = \ln(1 + u/r); \quad \varepsilon_r = \ln(1 + du/dr)$$

The final case which presents no problem even for large strains is the simple shear shown in Fig. 2.10 in which shear strain is traditionally defined as $\gamma = \tan\theta$.

We turn now to "infinitesimal strain" in which the strains are small enough for their products to be neglected. Considering the element shown in Fig. 2.11 where the deformation is greatly exaggerated, the normal strains $\varepsilon_x, \varepsilon_y, \varepsilon_z$ are defined using the "engineering strain" approach as the fractional change in the gauge lengths dx, dy, dz. The shear strains $\gamma_{xy}, \gamma_{yz}, \gamma_{zx}$ are defined by the angle by which the edges change from a right angle, that is, the same definition as used for simple shear in Fig. 2.10 with $\gamma = \tan\theta = \theta$ for infinitesimal strain.

Fig. 2.10 Simple shear

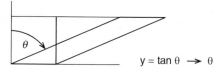

$y = \tan\theta \longrightarrow \theta$

Fig. 2.11 Element used for
analysis of strain

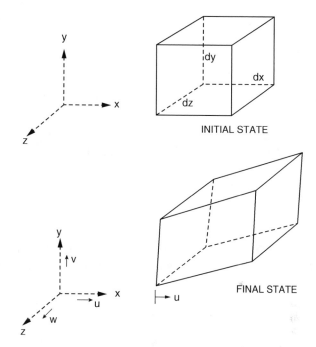

If we assume a continuous body, the six strain components are related to the
displacements (u, v, w) in the x, y, and z directions. This is illustrated in Fig. 2.12 for
the x–y plane. It is seen that

$$\varepsilon_x = \frac{\partial u}{\partial x}, \varepsilon_y = \frac{\partial v}{\partial y}, \gamma_{xy} = \frac{\partial u}{\partial y} + \frac{\partial v}{\partial x} \tag{2.6a}$$

and similarly

$$\varepsilon_z = \frac{\partial w}{\partial z}, \gamma_{yz} = \frac{\partial v}{\partial z} + \frac{\partial w}{\partial y}, \gamma_{zx} = \frac{\partial u}{\partial z} + \frac{\partial w}{\partial x} \tag{2.6b}$$

Just as for shear stresses, $\gamma_{xy} = \gamma_{yx}; \gamma_{yz} = \gamma_{zy}; \gamma_{zx} = \gamma_{xz}$ and a direction convention for the subscripts is unnecessary. Actually, this traditional engineering defini-
tion of shear strain is unfortunate for half of the shear strain as defined corresponds

Fig. 2.12 Relation of strains to displacement in x–y plane

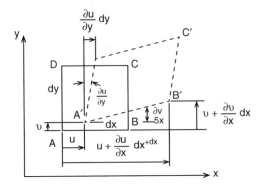

to a solid body rotation. This is illustrated schematically in Fig. 2.13. As a result in equations for determining principal strains or in transferring from one coordinate system to another, we have to use half of the shear strains as defined earlier. For example, the Mohr construction may be applied to situations in which $\varepsilon_x, \varepsilon_y, \gamma_{xy}$ (and $\gamma_{yz} = \gamma_{zx} = 0$) to determine say the principal strains $\varepsilon_1, \varepsilon_2$ and their direction relative to the x and y axes. However, instead of plotting shear strain as the ordinate, half of this value is taken as illustrated in Fig. 2.13.

Again, as with the stresses, it may be shown that the sum of the normal strains is independent of the orientation of the x, y, and z axes. Hence,

$$\varepsilon_x + \varepsilon_y + \varepsilon_y = \varepsilon_1 + \varepsilon_2 + \varepsilon_3 = 3\varepsilon_{av}$$

Also a deviatoric strain may be obtained by subtracting the average strain ε_{av} from the normal components of strain:

$$\varepsilon_x' = \varepsilon_x - \varepsilon_{av}; \varepsilon_y' = \varepsilon_y - \varepsilon_{av} \tag{2.7}$$

Since the six components of strain depend on only three displacements u, v, and w, they are not independent. Assuming that the displacements are single-valued continuous functions, from the definition of infinitesimal strains in terms of displacements, one obtains

$$\frac{\partial^2 \varepsilon_x}{\partial y^2} + \frac{\partial^2 \varepsilon_y}{\partial x^2} = \frac{\partial^2 \gamma_{xy}}{\partial x \partial y}; \quad \frac{2\partial^2 \varepsilon_x}{\partial y \partial z} = \frac{\partial}{\partial x}\left(\frac{\partial \gamma_{yz}}{\partial x} + \frac{\partial \gamma_{xz}}{\partial y} + \frac{\partial \gamma_{xy}}{\partial z}\right) \tag{2.8a}$$

and four similar equations by permuting subscripts. The strains must satisfy these "compatibility equations" and also any displacement boundary conditions imposed

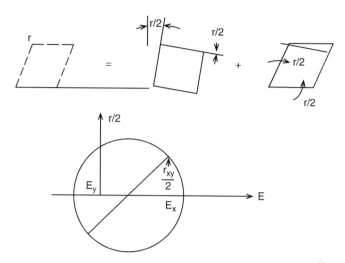

Fig. 2.13 Mohr circle for strain and justification for use of $\gamma/2$

on the body. Fortunately for the two-dimensional state of stress discussed in connection with the Mohr circle, the first of Eq. (2.8a) is the only compatibility equation involved. In polar coordinates, the radial and circumferential strains for axisymmetric deformation were seen to be functions only of the radial displacement u. This can be shown to impose the compatibility equation

$$\frac{r d\varepsilon_\theta}{dr} = \varepsilon_r - \varepsilon_\theta \tag{2.8b}$$

which applies also to a sphere loaded symmetrically (i.e., by internal or external pressure).

2.4 Requirements for a Solution

In general the calculation of stress and strain for any type of material behavior involves three aspects:

A. A state of stress which satisfies equilibrium and any loading boundary conditions
B. A state of strain which satisfies compatibility and any displacement boundary conditions
C. A material law sometimes called the constitutive equation, which relates stresses to deformations either through strain, rate of strain, or both, and possibly involving temperature and other environmental conditions

Later, in Sect. 2.10 we will review some of the methods of solution which have been used to satisfy these requirements. As an alternative, it is often possible to obtain bounds on the true solution, using variational methods when either aspect A or B listed above is neglected. In a few simple "statically determinate" cases (e.g., the thin-walled cylinder or sphere loaded by internal pressure), the stresses are obtained directly from equilibrium considerations and are independent of material behavior.

Before proceeding to discuss stress–strain relations, it may be useful to comment on the degree of uncertainty involved in describing the material law. In polycrystalline materials, the elastic constants are insensitive to small variations in alloying ingredients or microstructure as long as phase changes do not occur. For this reason, the equations of elasticity normally provide a precise description of material behavior. As a result, detailed analytical or numerical computations are justified since they are not limited by uncertainty as to material behavior. However, when we consider yield and subsequent plastic deformation, the material behavior cannot be defined as precisely. Plastic deformation depends on the history of loading, and the stress–strain curve beyond the elastic range may be strongly influenced by changes in chemical composition, microstructure, prior deformation, strain rate, and temperature. Because of these limitations, and the fact that the equations describing plastic deformation are less tractable than those for elasticity, considerable attention has been given to the development of approximate solutions for plasticity.

In turning now to discuss elastic and plastic deformation, it will be clearer, although somewhat cumbersome, to use the superscript notation ε^e and ε^p to denote elastic and plastic components of strain with ε being reserved for the total strain.

2.5   Elasticity

For elastic behavior and small strains, the material law in uniaxial tension ($\sigma_1 = \sigma$, $\sigma_2 = \sigma_3 = 0$) is the familiar Hooke's law $\sigma_1 = E\varepsilon_1^e$ which relates stress and the elastic component of strain through Young's Modulus, E. The transverse strains are given by $\varepsilon_2^e = \varepsilon_3^e = -\nu\varepsilon_1^e$ where ν is Poisson's ratio. For isotropic elastic behavior, there are only two independent elastic constants. Other constants which are used are the shear, or rigidity modulus, G, the ratio of shear stress to shear strain, and the bulk modulus, K, the ratio of hydrostatic stress to volumetric strain. That is, $K = p/3\varepsilon_{av}$ using the notation of Eqs. (2.3) and (2.7). It may be shown that

$$E = 2G(1 + \nu) \quad \text{and} \quad E = K(1 - 2\nu)$$

Another useful constant if stress is to be written in terms of strain is

$$\lambda = (E\nu)/(1 + \nu)(1 - 2\nu)$$

Often λ and G are referred to as Lame's constants.

Assuming that the strain components for multiaxial stresses are linear functions of the stresses and do not depend on the history of loading, Hooke's law may be generalized for isotropic multiaxial behavior to

$$\varepsilon_x^e = \frac{1}{E}\left[\sigma_x - \nu\left(\sigma_y + \sigma_z\right)\right]; \quad \gamma_{xy}^e = \frac{\tau_{xy}}{G}$$

$$\varepsilon_y^e = \frac{1}{E}\left[\sigma_y - \nu\left(\sigma_z + \sigma_x\right)\right]; \quad \gamma_{yz}^e = \frac{\tau_{yz}}{G} \qquad (2.9a)$$

$$\varepsilon_z^e = \frac{1}{E}\left[\sigma_z - \nu\left(\sigma_x + \sigma_y\right)\right]; \quad \gamma_{zx}^e = \frac{\tau_{zx}}{G}$$

Alternatively, using the definitions $\sigma_x' = \sigma_x - p;\ \varepsilon_x'^e = \varepsilon_x^e - \varepsilon_{av}^e$ the equations for $\varepsilon_x^e, \varepsilon_y^e, \varepsilon_z^e$ may be rewritten as

$$\varepsilon_x'^e = \frac{\sigma_x'}{2G};\, \varepsilon_y^e = \frac{\sigma_y'}{2G},\, \varepsilon_z^e = \frac{\sigma_z'}{2G} \qquad (2.9b)$$

If stress is to be written in terms of strain,

$$\sigma_x = (\lambda + 2G)\varepsilon_x^e + \lambda\varepsilon_y^e + \lambda\varepsilon_z^e \qquad (2.9c)$$

with two similar equations for σ_y and σ_z.

In our subsequent discussion of fracture and in surveying approaches to solution in Sect. 2.10, we make extensive use of energy methods. At this stage some of the basic definitions that are required will be summarized.

The work done in deforming unit volume of a homogeneous isotropic material when increments of total strain $d\varepsilon_x$, etc., occur in conjunction with stresses σ_x, etc., is

$$dW_1 = \sigma_x d\varepsilon_x + \sigma_y d\varepsilon_y + \sigma_z d\varepsilon_z + \tau_{xy}d\gamma_{xy} + \tau_{yz}d\gamma_{yz} + \tau_{zx}d\gamma_{zx} \qquad (2.10a)$$

Similarly, if stresses change while strains are held constant, we can define a quantity

$$dW_2 = \varepsilon_x d\sigma_x + \varepsilon_y d\sigma_y + \varepsilon_z d\sigma_z + \gamma_{xy}d\tau_{xy} + \gamma_{yz}d\tau_{yz} + \gamma_{zx}d\tau_{zx} \qquad (2.10b)$$

These expressions apply for any type of material behavior. However, for elastic behavior, the work done is stored in the material, and $dW_1 = dU_e$, the elastic strain energy per unit volume. The corresponding quantity $dV_2 = dU_c$ is sometimes referred to as the complementary energy, but it will be clearer if we use the term "complementary strain energy." The strain energy U_e' or complementary strain energy U_c' per unit volume corresponding to a final state of stress and strain can be obtained by integrating Eqs. (2.10a) and (2.10b) for the prescribed elastic stress–strain relation. For linear elastic behavior,

$$U'_e = U'_c = \frac{1}{2}\left[\sigma_x\varepsilon_x + \sigma_y\varepsilon_y + \sigma_z\varepsilon_z + \tau_{xy}\gamma_{xy} + \tau_{yz}\gamma_{yz} + \tau_{zx}\gamma_{zx}\right] \tag{2.10c}$$

The strain energy U_e or the complementary energy U_c for a body is obtained by integrating the expressions for U'_e and U'_c over the entire volume:

$$U_e = \iiint U'_e dxdydz; \quad U_c = \iiint U'_c dxdydz \tag{2.11}$$

Similar expressions for the strain energy can be written in other coordinate systems. Also U_e and U_c can be computed for a prescribed nonlinear elastic stress–strain relation. In this case they will no longer be equal.

2.6 The Yield Condition

As a ductile metal is loaded in a tension test, a stage is reached at which the deformation no longer recovers completely. The stress at which a significant amount of non-recoverable (plastic) deformation is observed is called the yield stress. It is important to realize that the definition of yield stress is quite arbitrary, and at least four distinct definitions have been used in the literature.

These are shown in Fig. 2.14 and are:

1. "0.002 offset in strain from the elastic line." This is the most common in engineering practice and the one used unless otherwise stated. Alternatively, another strain value may be used.

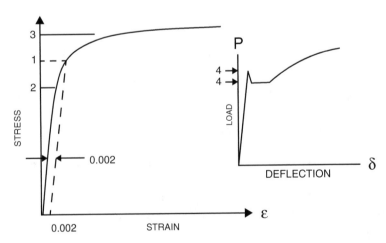

Fig. 2.14 Different definitions of yield. (1) 0.002 offset, (2) departure from linearity, (3) backward extrapolation, (4) upper or lower yield

Fig. 2.15 Load-deflection
record and effect of
unloading during tension
test

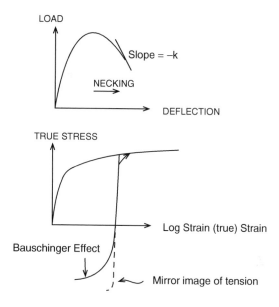

2. "Departure from linearity." This is sometimes used in research studies and calls
   for a definition of departure.
3. "Backward extrapolation."
4. The upper or lower yield point in materials which show this behavior.

An important problem is the prediction of yield under multiaxial stresses given
the value for uniaxial tension or some other materials test such as compression or
torsion. Before discussing this topic, it may be worth describing some aspects of the
tensile test.

The results of a uniaxial tension test are expressed most directly as a plot of load
versus deflection over a specified gauge length as shown in Fig. 2.15. Since the
specimen and testing machine can be viewed as springs in series, the test terminates
catastrophically when the slope of the curve equals $-k$, where k is the stiffness of
the testing machine. Fracture may precede this point if the material is brittle enough
in the tension test or may be prolonged to larger strains in a ductile metal, if a servo-
controlled machine, with essentially infinite stiffness, is used. Load can be
converted to "engineering stress" (load/original cross-sectional area A_0) or "true
stress" (load divided by current cross-sectional area A). Curves of engineering
stress–engineering strain or true stress–true strain can then be plotted. However,
after the maximum load is reached, deformation becomes localized with the
formation of a neck. Displacements based on a gauge length then have very little
meaning. Rather, the diameter of the neck has to be measured. Neglecting the
elastic components of strain which are usually very small compared to the plastic
components at the time of necking, the condition that volume is constant in plastic
deformation leads to $\frac{\pi}{4}d_0^2 L_0 = \frac{\pi}{4}d^2 L$ where L_0 is the original and L the final length of

a small gauge length at the neck, and d_0, d are the corresponding diameters. Thus, $\varepsilon = \ln\frac{L}{L_0} = 2\ln\frac{d_0}{d}$. After necking occurs, the state of stress is no longer uniaxial tension. An approximate solution due to Bridgeman [2], discussed in Chap. 3, allows a correction to be made for the triaxial stress state at the neck, and a true stress–true (logarithmic) strain curve can be obtained. With this information available, it is then possible to predict strains for other states of stress and to predict plastic instability (maximum load-carrying capacity) for pressure vessels and sheets carrying tensile loads.

It bears repeating that increased strain rate and decreased temperature elevate the yield stress, particularly in ordinary low-carbon steels. If a state of pure hydrostatic stress $\sigma_1 = \sigma_2 = \sigma_3$ is added to the tensile or any other test, it has insignificant effect on yield or the subsequent shape of the stress–strain curve but does influence the strain at fracture, increasing it for hydrostatic compression and decreasing it for hydrostatic tension. For example, with large enough superimposed hydrostatic pressure, cast iron can be extruded to form shapes such as a twist drill. The fact that the stress state under an indenter is very largely a triaxial compression accounts for the "plastic" behavior shown by glass when scratched or indented at low loads. Heat treatment, grain size, and minor alloying additions may greatly change the yield stress and stress–strain curve in steels and other metals while leaving elastic behavior relatively unchanged. A final aspect is the effect of "cold work" (plastic deformation) on the stress–strain curve. After an initial loading and unloading as shown in Fig. 2.15, reloading in the same direction essentially follows the curve obtained by monotonic loading. There may be a small hysteresis loop and some rounding of the corner at the yield point. One important exception to this is that, if unloaded and reloaded immediately, the upper yield does not appear. However, with aging or heating after the initial loading, the upper yield may reappear. Returning to Fig. 2.15b, when the direction of loading is reversed, the resulting yield stress may be considerably less than the stress reached in the initial loading. This phenomenon is referred to as the "Bauschinger effect." In a situation where cyclic loading occurs in the plastic range, one might expect strain hardening to lead to purely elastic behavior. However, instead a stable hysteresis loop develops after a few cycles. This could be thought of as another manifestation of the Bauschinger effect. Our subsequent discussion of yield and plastic deformation will be based on homogeneous isotropic material and will not consider the Bauschinger effect. If anisotropy develops as a result of plastic deformation, or is present initially, and if reversed loading or cyclic plastic deformation are involved, predictions become much more difficult and are beyond the scope of this introductory treatment.

Having measured the yield stress in a "simple" test such as tension or torsion, the next problem is one of predicting the combination of the principal stresses $(\sigma_1, \sigma_2, \sigma_3)$ at a point which determine whether yield occurs. Since yield and subsequent plastic deformation in metals occur by shear and are thus uninfluenced by adding a hydrostatic state of stress $(\sigma_1 = \sigma_2 = \sigma_3)$, the yield condition should involve only the principal shear stresses:

$$\frac{\sigma_1 - \sigma_2}{2}, \frac{\sigma_2 - \sigma_3}{2}, \frac{\sigma_3 - \sigma_1}{2}$$

The simplest approach in predicting yielding is to assume that only the largest of these shear stresses is important and that yield will occur when it reaches a critical value. If $\sigma_1 > \sigma_2 > \sigma_3$, this means that yield occurs when $\frac{\sigma_1 - \sigma_3}{2} = \pm$constant. An experimental measurement of yield is needed to establish the constant. Most often a tension test is used in which case $\sigma_1 = S_y$, $\sigma_2 = \sigma_3 = 0$ and the yield condition becomes $\sigma_1 - \sigma_3 = \pm S_y$. However, there is nothing unique about the tension test, and an experiment with other principal stress ratios could be used to obtain the constant in the yield condition. For example, in a torsion test on a thin-walled tube, the principal stresses are $\sigma_1 = -\sigma_3 = \tau$ (the shear stress), $\sigma_2 = 0$. If yield is observed at a shear stress $\tau = k$, then the yield criterion, which is associated with the French engineer Tresca[2] (1864), implies $2k = S_y$. If the labeling of the principal stress directions is arbitrary, then the maximum shear yield condition is written as

$$\text{Max}\{|\sigma_1 - \sigma_2|, |\sigma_2 - \sigma_3|, |\sigma_3 - \sigma_1|\} = S_y \text{ or } 2k \qquad (2.12)$$

That is, the maximum of the absolute values of the three quantities in parentheses is set equal to S_y or $2k$.

Alternatively, it can be argued that it is more realistic to include all the principal shear stresses in the yield criterion. In the absence of the Bauschinger effect, yield should occur at the same stress in tension as compression. Also, in a more general sense, the yield condition should be unchanged if the signs of all the principal stresses are reversed (i.e., $\sigma_1, \sigma_2, \sigma_3$ changed to $-\sigma_1, -\sigma_2, -\sigma_3$). This suggests that the yield criterion involves the squares of the shear stresses:

$$\left(\frac{\sigma_1 - \sigma_2}{2}\right)^2, \left(\frac{\sigma_2 - \sigma_1}{2}\right)^2 \text{ and } \left(\frac{\sigma_3 - \sigma_1}{2}\right)^2$$

The simplest way to combine these quantities is to add them which leads to the intuitive yield criterion:

$$\left(\frac{\sigma_1 - \sigma_2}{2}\right)^2 + \left(\frac{\sigma_2 - \sigma_1}{2}\right)^2 + \left(\frac{\sigma_3 - \sigma_1}{2}\right)^2 = \text{constant}$$

Again the constant is obtained from experiment.

For tension $(\sigma_1 = S_y, \sigma_2 = \sigma_3 = 0)$, the constant is $S_y^2/2$, and so the preceding equation becomes

---

[2] The development of yield criteria and the equations of plasticity are treated in many texts. For example, Hill [3] gives a thorough discussion with references to the papers of Tresca and later workers in this field that we will mention.

$$\frac{1}{\sqrt{2}}\left\{(\sigma_1 - \sigma_2)^2 + (\sigma_2 - \sigma_3)^2 + (\sigma_3 - \sigma_1)^2\right\}^{1/2} = S_y \qquad (2.13)$$

The left-hand side of this equation is essentially a "root mean square" shear stress which we will refer to as $\bar{\sigma}$ the "effective stress." If the torsion test is used to determine yield stress, inserting $\sigma_1 = -\sigma_3 = \kappa, \sigma_2 = 0$ leads to $\bar{\sigma} = \sqrt{3}\kappa$. Thus, the choice of the effective stress as a yield criterion implies $S_y = \sqrt{3}\kappa$.

The use of the "effective stress" to predict yield was first proposed by Maxwell (1856) in a letter to Kelvin, using essentially the reasoning presented here. Later, in 1913, the same expression was suggested by von Mises. His derivation is of interest in that it shows the types of functions of $\sigma_1, \sigma_2, \sigma_3$ which are acceptable yield criteria.

In discussing the analysis of stress, we saw that the three principal stresses were the solution of the cubic equation:

$$S^3 - I_1 S^2 - I_2 S - I_3 = 0$$

The yield condition for an isotropic material must be a symmetric function of the principal stresses. We now invoke that a theorem of algebra "A symmetric function of n variables" may be expressed in terms of n linearly independent symmetric functions of these n variables. Since I_1, I_2, I_3 are linearly independent symmetric functions of the principal stresses, this means that the yield condition may be written as $f(I_1, I_2, I_3) = $ constant.

This result is too general to be of much value. However, as the hydrostatic component of stress has no influence on yield, the stresses may be replaced by the stress deviators and the yield condition written as $f(J_1, J_2, J_3) = $ constant. If the yield condition is unchanged by changing the sign of all stresses, then the cubic term J_3 must enter as J_3^2, and, by definition, $J_1 = 0$. Hence, any yield criterion, which assumes isotropy, equivalence of tension and compression, and no effect due to the hydrostatic component of the stress state, may be written as

$$f\left(J_2, J_3^2\right) = \text{constant}$$

The simplest function satisfying this requirement is obtained by neglecting the role of J_3 and taking $J_2 = $ constant as the yield criterion. After some manipulation, it can be shown that $J_2 = \bar{\sigma}^2/3$, so the "effective stress" criterion could be regarded as the simplest function which satisfies the physical requirements.

The maximum shear yield condition must also be expressable as a function of J_2 and J_3^2. The result, which is more of general interest than for practical use, is given by Prager and Hodge [4] as

$$4J_2^3 - 27J_3^2 - 36\kappa^2 J_2^3 + 96\kappa^4 J_2 - 64\kappa^6 = 0$$

The quantity J_3 may be shown to be

$$\frac{1}{27}[(2\sigma_1 - \sigma_2 - \sigma_3)(2\sigma_2 - \sigma_3 - \sigma_1)(2\sigma_3 - \sigma_1 - \sigma_2)]$$

Various attempts have been made to give the effective stress yield criterion some additional physical significance but none of these are very convincing. One approach attempts to relate yielding to stored energy. Clearly the total strain energy cannot be used because hydrostatic stress will not produce yield. However, it was suggested by Hencky that the strain energy of distortion be taken as the criterion of yielding. By taking this "distortion energy" for a triaxial state of stress $\sigma_1, \sigma_2, \sigma_3$ and equating it to that in a tension specimen at yield, he found

$$\bar{\sigma} = S_y$$

Another suggestion is that yield occurs when the shear stress on a plane, which is inclined at an equal angle to the three principal directions, reaches a critical value. The shear stress on this so-called octahedral plane is merely $(\sqrt{2}/3)\bar{\sigma}$, so this octahedral stress approach is the same as the effective stress yield condition and is based on the suggestion that some statistical average of shear stress be taken over all possible orientations in a polycrystalline aggregate.

It is curious that the effective stress yield criterion can be derived from several different assumptions. However, this has no bearing on its validity which depends only on the success with which it correlates experimental results. Generally, it is found to provide predictions which are in good agreement with tests on ductile metals [3]. On the other hand, the predictions of the maximum shear yield criterion differ at most by 15.5 % from those of the effective stress. Thus, from a practical point of view, it matters very little which of the two criteria are used, and the choice can be based on convenience.

A number of yield criteria of the form $f(J_2^2, J_3^2) = $ constant have been proposed to improve the predictions of the simple $f(J^2) = $ constant ("effective stress") criterion. However, the arbitrariness of the definition of yield stress and its sensitivity to testing variables (strain rate, temperature) as well as to metallurgical variables (grain size, chemical composition, heat treatment) suggests that such refinements are rarely justified for practical problems.

If the principal stresses are not known, the effective stress may be found from

$$\bar{\sigma} = \frac{1}{\sqrt{2}}\left\{(\sigma_x - \sigma_y)^2 + (\sigma_y - \sigma_z)^2 + (\sigma_z - \sigma_x)^2 + 6\left(\tau_{xy}^2 + \tau_{yz}^2 + \tau_{zx}^2\right)\right\}^{1/2} \quad (2.14)$$

For example, for combined tension σ and shear stress τ,

$$\bar{\sigma} = \left(\sigma^2 + 3\tau^2\right)^{1/2}$$

while the maximum shear stress τ_{max} for the same loading is given by

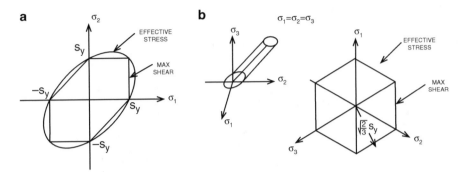

Fig. 2.16 (**a**) Comparison of yield criteria in two dimensions based on tensile yield value. (**b**) Yield surface in three dimensions based on tensile yield value. The normal to the plane shown is the axis of pure hydrostatic stress $\sigma_1 = \sigma_2 = \sigma_3$

$$\tau_{max} = \frac{1}{2}\left(\sigma^2 + 4\tau^2\right)^{1/2}$$

For the common case of two-dimensional states of stress with $\sigma_3 = 0$ and σ_1, σ_2 having arbitrary values, the two yield criteria are

$$\bar{\sigma} = \left(\sigma_1^2 - \sigma_1\sigma_2 + 3\sigma_2^2\right)^{1/2} = S_y \tag{2.15a}$$

$$\text{Max}\{|\sigma_1 - \sigma_2|, |\sigma_2|, |\sigma_3|\} = S_y \tag{2.15b}$$

These equations, in which the tensile test is used to establish the yield strength, are discussed in texts on Strength of Materials and Machine Design. In graphical form the two yield criteria are shown as the ellipse, Eq. (2.15a), and inscribed hexagon shown in Fig. 2.16a. These represent two dimensional cuts of the yield criteria plotted in three dimensions; here they become a circular cylinder and the inscribed regular hexagonal cylinder shown in Fig. 2.16b. The axes of these cylinders correspond to the hydrostatic state of stress $\sigma_1 = \sigma_2 = \sigma_3$. The terms "yield surface" or "yield locus" are often used to describe these graphical representations of the yield conditions.

Since the tensile test is used so often to establish the constant in the yield criterion, it is worth recalling that any other test could be used, such as torsion of a thin-walled tube. In this case, the yield surfaces would have to coincide for the case of pure torsion. The maximum shear condition would then be represented by a hexagonal cylinder circumscribing the circular cylinder of the effective stress criterion.

2.7   Plasticity

On unloading after plastic deformation has taken place and then reloading in the same direction, the value of the effective stress (or maximum shear stress) at which yield occurs will increase. This effect is called "strain hardening" and would be expected to be a function of the plastic components of strain. Since plastic deformation occurs by shear, it is reasonable to expect the new value of the effective stress at yield to depend on the principal plastic shear strains:

$$\bar{\sigma} = f\left(\varepsilon_1^P - \varepsilon_2^P, \varepsilon_2^P - \varepsilon_3^P, \varepsilon_3^P - \varepsilon_1^P\right)$$

By analogy to the effective stress, an effective plastic strain is defined as

$$\bar{\varepsilon}^P = \frac{\sqrt{3}}{2}\left\{\left(\varepsilon_1^P - \varepsilon_2^P\right)^2 + \left(\varepsilon_2^P - \varepsilon_3^P\right)^2 + \left(\varepsilon_3^P - \varepsilon_1^P\right)^2\right\}^{1/2} \tag{2.16a}$$

or alternatively in terms of the x, y, and z components of strain

$$\bar{\varepsilon}^P = \frac{\sqrt{2}}{2}\left\{\left(\varepsilon_x^P - \varepsilon_y^P\right)^2 + \left(\varepsilon_y^P - \varepsilon_z^P\right)^2 + \left(\varepsilon_z^P - \varepsilon_x^P\right)^2 + \frac{3}{2}\left(\gamma_{xy}^{P2} + \gamma_{yz}^{P2} + \gamma_{zx}^{P2}\right)\right\}^{1/2} \tag{2.16b}$$

and the assumption is made that

$$\bar{\sigma} = f(\bar{\varepsilon}^P) \tag{2.17}$$

The numerical quantities $\left(\sqrt{2}/3\right)$ in the definition of $\bar{\varepsilon}^P$ and $\left(1/\sqrt{2}\right)$ in the definition of $\bar{\sigma}$ are chosen, for convenience, so that $\bar{\sigma}$ and $\bar{\varepsilon}^P$ reduce to the tensile stress σ_1 and the axial plastic strain ε_1^P in a uniaxial tension test. Thus, tensile test results plotted as σ_1 versus ε_1^P may merely be relabeled $\bar{\sigma}$, $\bar{\varepsilon}^P$ to obtain a curve of effective stress versus effective plastic strain. To examine the validity of Eq. (2.17), $\bar{\sigma}$ and $\bar{\varepsilon}^P$ can be obtained from several different tests. For example:

(a) Torsion of a thin-walled tube with shear stress τ and plastic shear strain γ^P

$$\sigma_1 = -\sigma_3 = \tau, \sigma_2 = 0; \bar{\sigma} = \sqrt{3}\tau$$
$$\varepsilon_1^P = -\varepsilon_3^P = \frac{\gamma^P}{2}, \varepsilon_2^P = 0; \bar{\varepsilon}^P = \gamma^P/\sqrt{3}$$

(b) Internal pressurization of a thin-walled tube with closed ends, inside radius R, thickness h, pressure p

$$\sigma_\theta = \frac{pR}{h}, \quad \sigma_a = \frac{pR}{2h}, \quad \sigma_r \approx 0; \quad \bar{\sigma} = \frac{\sqrt{3}}{2}\sigma_\theta$$

It is seen later, from Eqs. (2.22a) and (2.22b), that $\varepsilon_a^P = 0$; hence

$$\varepsilon_\theta^{\mathrm{p}} = -\varepsilon_r^{\mathrm{p}}; \quad \bar{\varepsilon}^{\mathrm{p}} = \frac{2}{\sqrt{3}}\,\varepsilon_\theta^{\mathrm{p}}$$

(c) Internal pressurization of a thin-walled sphere, inside radius R, thickness h, pressure p

$$\sigma_\theta = \sigma_\phi = \frac{pR}{2h}, \quad \sigma_r \approx 0; \quad \bar{\sigma} = \sigma_\theta$$
$$\varepsilon_\theta^{\mathrm{p}} = \varepsilon_\phi^{\mathrm{p}} \text{ by symmetry,} \quad \text{hence } \varepsilon_r^{\mathrm{p}} = -2\varepsilon_\theta^{\mathrm{p}}; \quad \bar{\varepsilon}^{\mathrm{p}} = 2\varepsilon_\theta^{\mathrm{p}}$$

At this stage we note that the equivalence of tension and compression stress–strain curves follows if Eq. (2.17) correctly describes material behavior and the logarithmic strain is used. Experiments on annealed isotropic ductile metals using the preceding tests give very similar curves on a plot of $\bar{\sigma}$ versus $\bar{\varepsilon}^{\mathrm{p}}$ if the strains are not too large. This is illustrated by test results shown in Fig. 2.17.[3] It is interesting that the separation of the tension and torsion curves at strains $\bar{\varepsilon}^{\mathrm{p}} > 0.2$ agrees with the value quoted by Hill [3], who suggests reasons for the effect. There is thus an additional source of uncertainty if plasticity analyses are carried out to large strains.

As an alternative to Eqs. (2.15a) and (2.15b), stresses and plastic strains may be related through the maximum shear stress and maximum plastic shear strain. That is,

$$\tau_{\max} = f\left(\gamma_{\max}^{\mathrm{p}}\right) \tag{2.18}$$

For example, for the tension test, $\tau_{\max} = \sigma/2$, and since $\varepsilon_2^{\mathrm{p}} = \varepsilon_3^{\mathrm{p}} = -\frac{1}{2}\varepsilon_1^{\mathrm{p}}, \gamma_{\max}^{\mathrm{p}} = \frac{3}{2}\varepsilon_1^{\mathrm{p}}$. Agreement between the different tests is usually not as close for small strains as when Eq. (2.17) is used. This is seen in Fig. 2.18 which is a replot of the tension and torsion curves of Fig. 2.17. Perhaps fortuitously, Eq. (2.18) provides a better correlation between the two tests at larger strains.

Our preceding discussion of plastic deformation has been oversimplified because unlike elastic deformation, it is path dependent. That is, at the end of a prescribed loading sequence, the final values of stress cannot, in general, be related to the final plastic components of strain. This can be illustrated by visualizing an experiment in a specimen that is first deformed well beyond the yield in tension and then compressed to its initial shape. The final strains are zero, but physical experience with metals tells us that the material will have been strain hardened. On the other hand, if the material is an amorphous polymer, its original properties may have been restored. At least for metals, this experiment suggests that some measure of effective plastic strain must be used which does not decrease when the direction

---

[3] Instability, comparable to necking in the tension test, limits the usefulness of pressurized specimens, and thin-walled tubes loaded in torsion will buckle at a certain shear strain. Fortunately, it is possible to obtain shear stress–shear strain data from solid circular rods in torsion.

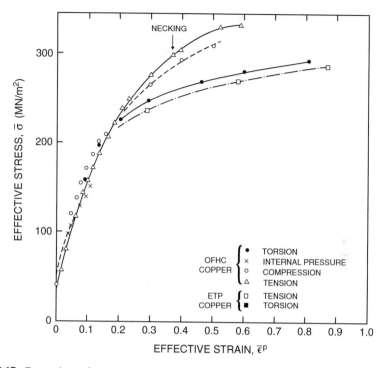

Fig. 2.17 Comparison of tension, compression, torsion, and internal pressure test results on the basis of the effective stress–effective plastic strain relations. Annealed OFHC copper and annealed ETP copper

of deformation is reversed. This can be done by looking at the deformation incrementally and writing

$$\Delta\bar{\varepsilon}^P = +\frac{\sqrt{3}}{2}\left\{\left(\Delta\varepsilon_1^P - \Delta\varepsilon_2^P\right)^2 + \left(\Delta\varepsilon_2^P - \Delta\varepsilon_3^P\right)^2 + \left(\Delta\varepsilon_3^P - \Delta\varepsilon_1^P\right)^2\right\}^{1/2} \quad (2.19)$$

where the + sign emphasizes the fact that $\Delta\bar{\varepsilon}^P$, like $\bar{\sigma}$ and $\bar{\varepsilon}^P$, is always positive.
    Then, Eq. (2.17) may be replaced by

$$\bar{\sigma} = f\left(\sum \Delta\bar{\varepsilon}^P\right) \quad (2.20a)$$

In the limit, replacing increments by differential quantities

$$\bar{\sigma} = f\left(\int d\bar{\varepsilon}^P\right) \quad (2.20b)$$

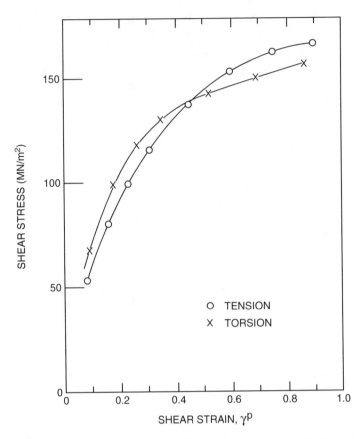

Fig. 2.18 Shear stress–plastic shear strain curves for tension and torsion tests. (Annealed OFHC copper)

where $d\bar{\varepsilon}^p$ is defined by Eq. (2.19) with $\Delta\bar{\varepsilon}^p$ replaced by $d\bar{\varepsilon}^p$, etc. For many practical situations, $\bar{\varepsilon}^p = \int d\bar{\varepsilon}^p$, and only the final plastic components of strain need to be considered. This occurs when:

1. There is no reversal of loading.
2. The principal stress ratios remain constant during deformation.
3. The principal stress directions remain fixed during deformation.

Even if some departure occurs from the last two requirements, it is often found that $\bar{\varepsilon}^p \approx \int d\bar{\varepsilon}^p$. However, if reversed loading occurs, as in the tension–compression experiment mentioned earlier, gross errors can result if the history of loading is ignored.

Equation (2.20b) implies that the yield surface expands isotropically as a result of plastic deformation. This is inconsistent with the Bauschinger effect, and many studies have been made to examine the effect of prior plastic deformation on the shape of the subsequent yield surface. Fortunately, many problems involve mono- tonic loading with fixed principal stress ratios and directions so this complication does not arise.

So far, only the effective plastic strain increment $d\bar{\varepsilon}^P$ can be predicted from tension data when the effective stress $\bar{\sigma}$ is increased. A rule is needed from which the individual components of plastic strain increment can be predicted. The traditional approach is to assume that the principal stress directions coincide with the direc- tions of the principal plastic strain increments and that the principal plastic shear strain increments are proportional to the principal shear stresses. That is,

$$\frac{d\varepsilon_1^P - d\varepsilon_2^P}{\sigma_1 - \sigma_2} = \frac{d\varepsilon_2^P - d\varepsilon_3^P}{\sigma_2 - \sigma_3} = \frac{d\varepsilon_3^P - d\varepsilon_1^P}{\sigma_3 - \sigma_1} = d\lambda \qquad (2.21)$$

where $d\lambda$ is constant at a given stage of the load history at a given point in the material. From Eq. (2.21), it may be shown that

$$\frac{d\varepsilon_1^P}{\sigma_1'} = \frac{d\varepsilon_x^P}{\sigma_x'} = \frac{d\gamma_{xy}^P}{2\tau_{xy}'}$$

with similar expressions relating the other components of plastic strain increment and deviatoric normal stresses or shear stress. Equation (2.21) is often referred to as the Levy–Mises equation. However, both Levy and Mises used increments of total strain rather than plastic strain, and the equation in its present form is due to Prandtl and Reuss.

An alternative and more recent derivation of Eq. (2.21) is perhaps more attrac- tive physically. It makes use of the "principle of maximum plastic resistance" which will be referred to again in Sect. 2.10.

This states: "For any plastic strain increment, the state of stress actually occur- ring gives an increment of work which equals or exceeds the work which would be done by that plastic strain increment with any other state of stress within or on the yield locus."

The proof of this principle for single crystals and aggregates of single crystals is due to Bishop and Hill [5]. The statement of the principle we have given is essentially that presented by McClintock and Argon in their comprehensive text. [6] The reader is referred to this reference for the demonstration that the "principle of maximum plastic resistance" requires that the plastic strain increment be normal to the yield surface. For the effective stress yield criterion, this leads to

$$\frac{d\varepsilon_1^P}{\partial\bar{\sigma}/\partial\sigma_1} = \frac{d\varepsilon_2^P}{\partial\bar{\sigma}/\partial\sigma_2} = \frac{d\varepsilon_3^P}{\partial\bar{\sigma}/\partial\sigma_3}$$

This result is identical to Eq. (2.21), which can now be interpreted as the flow rule associated with the choice of the effective stress as the yield criterion. Similarly, the flow rule associated with the maximum shear stress yield criterion is seen, for $\sigma_1 > \sigma_2 > \sigma_3$, to be $d\varepsilon_1^P = -d\varepsilon_3^P$; $d\varepsilon_2^P = 0$. Some ambiguity occurs at a corner on the yield surface, with one approach being to take $d\varepsilon_1^P = -d\varepsilon_2^P = -d\varepsilon_3^P/2$.

Whatever the interpretation we choose for Eq. (2.21), combining it with the definition of effective stress and effective plastic strain increment and the condition that volume is constant leads to

$$d\varepsilon_1^P = \frac{d\bar{\varepsilon}^P}{\bar{\sigma}}\left[\sigma_1 - \frac{1}{2}(\sigma_2 + \sigma_3)\right] \tag{2.22a}$$

$$\text{or } d\varepsilon_x^P = \frac{d\bar{\varepsilon}^P}{\bar{\sigma}}\left[\sigma_x - \frac{1}{2}(\sigma_y + \sigma_z)\right]; \quad d\gamma_{xy}^P = 3\frac{d\varepsilon^P}{\bar{\sigma}}\tau_{xy} \tag{2.22b}$$

and similar equations for $d\varepsilon_2^P, d\varepsilon_y^P$, etc. Thus, when $\bar{\sigma}$ increases by a small amount, $d\bar{\varepsilon}^P$ is obtained from say tensile data, and the components $d\varepsilon_1^P$, etc., can be calculated. If the three conditions listed earlier for the equivalence of $\bar{\varepsilon}^P$ and $\int d\bar{\varepsilon}^P$ are satisfied, the preceding equations may be written as

$$\varepsilon_1^P = \frac{\bar{\varepsilon}^P}{\bar{\sigma}}\left[\sigma_1 - \frac{1}{2}(\sigma_2 + \sigma_3)\right] \tag{2.23a}$$

$$\varepsilon_x^P = \frac{\bar{\varepsilon}^P}{\bar{\sigma}}\left[\sigma_x - \frac{1}{2}(\sigma_y + \sigma_z)\right]; \qquad \gamma_{xy}^P = 3\frac{\varepsilon^P}{\bar{\sigma}}\tau_{xy} \tag{2.23b}$$

Since $\bar{\varepsilon}^P$ is known as a function of $\bar{\sigma}$, from say a tension test, Eq. (2.23a), for example, may be rewritten as

$$\varepsilon_1^P = \frac{F(\bar{\sigma})}{\bar{\sigma}}\left[\sigma_1 - \frac{1}{2}(\sigma_2 + \sigma_3)\right] \quad \text{where} \quad \bar{\varepsilon}^P = F(\bar{\sigma})$$

These incremental and final plastic strain approaches are often referred to in the literature as "flow theory" and "deformation theory." While the deformation theory (final strain) equations are only exact in special cases, they are much simpler computationally than the incremental equations and have provided useful approximate solutions in many situations. By neglecting the path dependence of plastic deformation, the deformation theory equations are essentially equivalent to those for nonlinear elasticity with Poisson's ratio equal to one half. The flow theory predictions will also be in error when plastic strains change sign unless special precautions are taken to allow for the Bauschinger effect. However, more complicated constitutive equations and more extensive materials test data are required before this can be done.

Unless the plastic components of strain are large compared to the elastic components, as in metal forming, both components have to be considered. The resulting equations for incremental plasticity are known as the Prandtl–Reuss equations:

$$d\varepsilon_1 = \frac{1}{E}[d\sigma_1 - \nu(d\sigma_2 + d\sigma_3)] + \frac{d\bar{\varepsilon}^p}{\bar{\sigma}}\left[\sigma_1 - \frac{1}{2}(\sigma_2 + \sigma_3)\right] \qquad (2.24a)$$

$$\text{or } d\varepsilon_x = \frac{1}{E}\left[d\sigma_x - \nu(d\sigma_y + d\sigma_z)\right] + \frac{d\bar{\varepsilon}^p}{\bar{\sigma}}\left[\sigma_x - \frac{1}{2}(\sigma_y + \sigma_z)\right]$$

$$d\gamma_{xy} = \frac{d\tau_{xy}}{G} + 3\frac{d\bar{\varepsilon}^p}{\bar{\sigma}}\tau_{xy} \qquad (2.24b)$$

For the deformation theory of plasticity, the corresponding equations are

$$\varepsilon_1 = \frac{1}{E}[d\sigma_1 - \nu(d\sigma_2 + d\sigma_3)] + \frac{\bar{\varepsilon}^p}{\bar{\sigma}}\left[\sigma_1 - \frac{1}{2}(\sigma_2 + \sigma_3)\right] \qquad (2.25a)$$

$$\text{or } \varepsilon_x = \frac{1}{E}\left[\sigma_x - \nu(\sigma_y + \sigma_z)\right] + \frac{\bar{\varepsilon}^p}{\bar{\sigma}}\left[\sigma_x - \frac{1}{2}(\sigma_y + \sigma_z)\right]$$

$$\gamma_{xy} = \frac{\tau_{xy}}{G} + 3\frac{\bar{\varepsilon}^p}{\bar{\sigma}}\tau_{xy} \qquad (2.25b)$$

2.8 Creep and Viscous Behavior

At a sufficiently elevated temperature, solid materials will deform continuously with time under constant stress. We call this phenomenon "creep." In metals most of the creep strain is irrecoverable, and the strain components are predicted using the same equations as incremental plasticity with strain increments replaced by the creep strain rates, etc., where the dot denotes differentiation with respect to time. Thus,

$$\dot{\varepsilon}_1^c = \frac{\dot{\bar{\varepsilon}}^c}{\bar{\sigma}}\left[\sigma_1 - \frac{1}{2}(\sigma_2 + \sigma_3)\right] \qquad (2.26)$$

The effective creep stain rate $\dot{\bar{\varepsilon}}^c$ is defined in the same way as $\Delta\bar{\varepsilon}^p$ when $\Delta\varepsilon_1^p$ is replaced by $\dot{\varepsilon}_1^c$, etc. Various creep laws have been proposed to relate $\dot{\varepsilon}_1^c$ to $\bar{\sigma}$, and other variables and detailed descriptions can be found in the literature [7]. We mention only the simple case of steady-state creep which may often be described by $\dot{\varepsilon}^c = B\sigma^n$ in uniaxial tests and hence by $\dot{\bar{\varepsilon}}^c = B\bar{\sigma}^n$ for multiaxial states of stress. Typical values of n for low-alloy and austenitic steels in high-temperature service are 4–5, but exponents ranging from 1 to 8 or greater have been reported. For steady creep with $\dot{\varepsilon}^c = B\sigma^n$, the strain-rate equations have the simple form:

$$\dot{\varepsilon}_1^c = B(\bar{\sigma})^{n-1}\left[\sigma_1 - \frac{1}{2}(\sigma_2 + \sigma_3)\right] \qquad (2.27)$$

The special case of steady-state creep in which $n = 1$ is linearly viscous behavior. Usually the coefficient of viscosity η is defined from a shear test as $\tau = \eta d\gamma^c / dt$. Relating τ and $\dot{\gamma}^c$ to $\bar{\sigma}$, we find that

$$\dot{\varepsilon}_1^c = \frac{1}{3\eta}\left[\sigma_1 - \frac{1}{2}(\sigma_2 + \sigma_3)\right] \tag{2.28a}$$

$$\text{or } \dot{\varepsilon}_x^c = \frac{1}{3\eta}\left[\sigma_x - \frac{1}{2}(\sigma_y + \sigma_z)\right]$$
$$\dot{\gamma}_{xy}^c = \frac{\tau_{xy}}{\eta} \tag{2.28b}$$

As with Eqs. (2.23a) and (2.23b), the total strain rate is obtained by adding the elastic value $\dot{\varepsilon}_1^e = \frac{1}{E}\left[\dot{\sigma}_1 - \frac{1}{2}(\dot{\sigma}_2 + \dot{\sigma}_3)\right]$ to the creep or viscous strain rate. The combination of linearly viscous and elastic behavior (viscoelastic material) is useful in describing many polymeric materials and has received a great deal of attention in the literature. For elasticity problems in which the stresses are not functions of Poisson's ratio, the stress distribution for viscous or viscoelastic behavior will be the same as for elastic behavior.

2.9 Plane Stress and Plane Strain

A three-dimensional stress analysis which combines the equations of equilibrium, compatibility, and the stress–strain relations discussed in the past sections will usually require a numerical solution. Fortunately, a large number of cases involve a "plane" or two-dimensional problem which greatly simplifies the analysis and often allows analytical solutions to be obtained.

First, consider the flat plate shown in Fig. 2.19a which lies in the x–y plane. It has a uniform thickness in the z direction which is small compared to its dimensions in the x

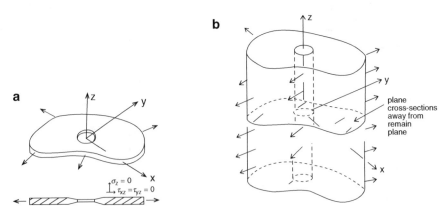

Fig. 2.19 (a) "Plane stress" with displacements in z direction at hole exaggerated. (b) "Plane strain" in z direction compared dimension in y plane

and y directions. Loading is applied only in the x–y plane and is uniformly distributed in the z direction. Since the top and bottom surfaces of the part are free of loading, the boundary conditions there are $\sigma_z = \tau_{zx} = \tau_{zy}$. In a thin plate, these stresses cannot reach appreciable values between the two faces and is assumed that they are zero everywhere. Also, with loading being uniform in the z direction, it is assumed that the remaining stress components σ_x, σ_y and τ_{xy} are not functions of z. These assumptions are often referred to as those of "generalized plane stress." They lead to solutions which while adequate in most problems are not rigorous three-dimensional solutions. The corrections required to the generalized plane stress solution as thickness increases are discussed for elastic behavior by Timoshenko and Goodier [8]. Here, we use the terminology "plane stress" and will discuss later the thickness limitations required to satisfy the assumed stress state in fracture problems. The strain in the z direction is a function of x and y, and since the plate is thin, it is assumed that there is no constraint to expansion or contraction in the z direction. Thus, as shown schematically in Fig. 2.19a, the plate will become nonuniform in thickness.

The other situation, that of "plane strain," involves a prismatic cylinder such as shown in Fig. 2.19b in which the z dimension is large compared to the x and y dimensions. Loading is again only in the x–y plane and is assumed the same at every cross section. Deformation in the z direction cannot occur as in the case of plane stress, or voids would open up. Thus, away from the ends, plane cross sections before loading will remain plane when load is applied and will have the same strains in the x–y plane. These assumptions of plane strain lead to $\varepsilon_z = $ constant (which may be a function of time) and $\gamma_{xz} = \gamma_{yz} = 0$. For isotropic behavior, for any of the types of material behavior that have been discussed, $\gamma_{xz} = \gamma_{yz} = 0$ also implies $\tau_{xz} = \tau_{yz} = 0$. So for plane strain, the stresses to be considered are σ_x, σ_y, and τ_{xy}, which are only functions of x and y, as well as σ_z which will depend on the values of ε_z, σ_x, and σ_y. Later, in discussing the stress fields around sharp cracks and in certain other solutions, a special case of plane strain will be invoked in which $\varepsilon_z = 0$ with consequent simplification in the expression for σ_z.

Plane strain cannot occur at or near a free surface, where $\sigma_z = 0$. In Chap. 3 in discussing solutions for members with holes or notches, some values will be given for the distance from the free surface at which the assumption of plane strain is satisfied.

For both plane stress and plane strain, the equilibrium equations (Eq. 2.4b) simplify to

$$\frac{\partial \sigma_x}{\partial x} + \frac{\partial \tau_{xy}}{\partial y} = 0 \qquad \frac{\partial \tau_{xy}}{\partial x} + \frac{\partial \sigma_x}{\partial y} = 0$$

and only one of the six equations expressing compatibility of strain (Eq. 2.8a) is required:

$$\frac{\partial^2 \varepsilon_x}{\partial y^2} + \frac{\partial^2 \varepsilon_y}{\partial x^2} = \frac{\partial^2 \gamma_{xy}}{\partial x \partial y}$$

For elastic behavior, the stress–strain relations may be written in the form

$$\text{Plane stress}: \quad 2G\varepsilon_x^e = \sigma_x - \frac{\nu}{1+\nu}(\sigma_x + \sigma_y)$$

$$2G\varepsilon_y^e = \sigma_y - \frac{\nu}{1+\nu}(\sigma_x + \sigma_y)$$

$$G\gamma_{xy}^e = \tau_{xy}$$

$$\text{Plane strain}: \quad 2G\varepsilon_x^e = \sigma_x - \nu(\sigma_x + \sigma_y)$$

$$2G\varepsilon_y^e = \sigma_y - \nu(\sigma_x + \sigma_y)$$

$$G\gamma_{xy}^e = \tau_{xy}$$

For plane strain if $\varepsilon_x^e = $ constant rather than $\varepsilon_x^e = 0$, $\sigma_z = E\varepsilon_x^e + \nu(\sigma_x + \sigma_y)$ and additional strains $\varepsilon_x^e = \varepsilon_y^e = \nu\varepsilon_z^e$ have to be added to those calculated for $\varepsilon_z^e = 0$. However, since these additional strains are not functions of x or y, they do not alter the compatibility equation. Thus, comparing the stress–strain relations for plane stress and plane strain, it is seen that solutions for plane stress can be applied to plane strain if ν is replaced by $\nu/(1-\nu)$.

Considering the case of plane stress, the compatibility equation may be written in terms of stresses as

$$\frac{\partial^2}{\partial y^2}(\sigma_x - \nu\sigma_y) + \frac{\partial^2}{\partial x^2}(\sigma_y - \nu\sigma_x) = 2(1+\nu)\frac{\partial^2 \tau_{xy}}{\partial x \partial y} \qquad (2.29)$$

The problem is now reduced to solving this equation and the two equilibrium equations for the three stresses.

The two equilibrium equations are automatically satisfied by defining a stress function ψ such that

$$\frac{\partial^2 \psi}{\partial y^2} = \sigma_x; \quad \frac{\partial^2 \psi}{\partial x^2} = \sigma_y; \quad \frac{\partial^2 \psi}{\partial x \partial y} = -\tau_{xy}$$

Using these expressions to substitute for the stresses in Eq. (2.29) leads to

$$\frac{\partial^4 \psi}{\partial x^4} + \frac{2\partial^4 \psi}{\partial x^2 \partial y^2} + \frac{\partial^4 \psi}{\partial y^4} = 0 \qquad (2.30)$$

This equation also applies to plane strain but does not include body forces or dynamic loading. Provided body forces are absent or their derivatives $\partial F_x/\partial x$, $\partial F_y/\partial y$ both equal to zero [8], the solution for the stresses does not contain the elastic constants and is the same for plane stress and plane strain if the shape of the cross section and the boundary loading per unit thickness are the same.

Equation (2.30) is usually written in terms of the Laplace, or harmonic, operator $\mathrm{DEL}^2 = \frac{\partial^2}{\partial x^2} + \frac{\partial^2}{\partial y^2}$ as

$$\nabla^4 \psi = 0 \tag{2.31}$$

and is referred to as the biharmonic equation.

Few solutions to this equation have been obtained by direct analytical means. More often, the method of solution is an inverse one in which a stress function is chosen, and the problem it solves is then obtained. Many solutions of this type are discussed in the literature (e.g., in [8]).

2.10   Methods of Solution

A large number of different analytical, numerical, and experimental methods have been used to obtain the solutions that we employ in studying fracture. When sharp cracks are involved, the large strain gradients near the crack tip present problems in numerical and experimental studies. For this reason, even in this "computer age," it is particularly important to have analytical solutions against which the accuracy of numerical or experimental results can be compared.

In a general three-dimensional problem involving linear elastic behavior, there are 15 equations in 15 unknowns (6 stresses, 6 strains, 3 displacements) which have to be satisfied as well as the boundary conditions. The 15 equations may be reduced to 3 second-order partial differential equations in the displacements u, v, and w, but even so, it is a formidable problem to determine functions which satisfy these equations and the boundary conditions. As a result there are relatively few analytical solutions for three-dimensional linear elasticity and even fewer for more complicated stress–strain laws. The majority of the analytical solutions that have been obtained for any type of material behavior are for the two-dimensional cases of plane stress and plane strain with some solutions also being available for axisymmetric problems.

Direct integration of the stress–strain equations is only feasible for the simplest problems. For example, the circular hole or inclusion subjected to uniform radial loading in plane stress or plane strain and the spherical hole or inclusion subjected to uniform radial loading may be studied analytically for elastic behavior, plastic behavior, or the combination of elasticity and non-strain-hardening plasticity. We will make us of some of these solutions in discussing ductile fracture by void growth in Chap. 7.

For problems involving plane stress or plane strain and elastic behavior, the stress function approach has been used extensively for both analytical and numerical solution. In some cases, simple algebraic or trigonometric expressions for the stress function enable solutions to be obtained. An example which we will discuss in the next chapter is the circular hole in a flat plate loaded by uniaxial tension. For more complicated problems involving elliptical holes or line cracks, it becomes advantageous for analytical studies to work with stress functions which are expressed in terms of complex variables. The most general formulation, which has been used to study many cracked configurations, is that of Muskhelishvili [9]. More restricted but considerably simpler approaches due to Westergaard [10] and Williams [11] are adequate for our purposes and will be outlined in the next chapter. A number of advanced analytical and numerical techniques have been used to obtain elastic solutions for parts containing cracks. A review of some of these is given in [12]. For numerical solution, the complex variable approach appears to have been of limited value, even before finite element methods were developed.

We mention only two approaches to direct numerical solution of the equations. One is "boundary collocation" in which a truncated series which satisfies the biharmonic equation in the interior of the body has its coefficients chosen to satisfy the boundary conditions at a selected number of points. For example, Srawley and Gross [13] have given solutions obtained by this technique for the two cracked specimens standardized by ASTM. This is only one of a variety of discretization methods, also called "methods of weighted residuals," which can be applied to the numerical solution of the biharmonic equation. Another approach is the solution of the biharmonic equation or the basic stress–strain relations in finite-difference form. For example, using the mesh of Fig. 2.20, we can approximate $\partial \psi / \partial x$ at mesh point i as the first central difference $(\partial \psi / \partial x)_i = (1/2h)(\psi_{i+1} - \psi_{i-1})$ where points $i+1$ and $i-1$ bound point i in the x coordinate direction. Similarly, the second difference becomes $\left(\partial^2 \psi / \partial^2 x\right)_i = (1/4h^2)(\psi_{i+2} - 2\psi_i + \psi_{i-2})$. In this way the biharmonic equation at point 0 can be approximated by a difference equation in the 12 surrounding points as

$$20\psi_0 - 8[\psi_1 + \psi_2 + \psi_3 + \psi_4] + 2[\psi_6 + \psi_8 + \psi_{10} + \psi_{12}] + \psi_5 + \psi_7 + \psi_9 + \psi_{11} = 0$$

Similar equations are written for each mesh point in the specimen. Solutions are sought for the values of ψ_i which satisfy these equations and the boundary conditions, the latter often being the most difficult aspect. This approach was pioneered by Southwell [14] who used manual "relaxation" methods to solve the "biharmonic" and other equations in numerical form. Non-strain-hardening plastic deformation may also be analyzed by this method of solution. As the simplest

Fig. 2.20 Mesh used to solve Biharmonic equation

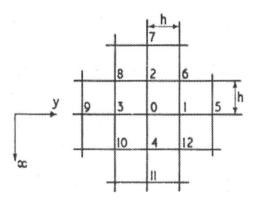

example, if one considers plane strain with incompressible behavior ($\nu = 0.5$) for both elastic and plastic components of strain in the yielded region, then $\sigma_z = (\sigma_x + \sigma_y)/2$, and the effective stress yield condition becomes

$$\left(\sigma_x - \sigma_y\right)^2 + 4\tau_{xy}^2 = \frac{4}{3}S_y^2$$

Since the left side of the equation is identically $\nabla^4\psi$, the solution of the elastic–plastic problem requires $\nabla^4\psi = 0$ in the elastic regions and $\Delta^4\psi = (4/3)S_y^2$ in yielded regions. This type of approach was used to obtain the first elastic–plastic solutions for parts containing notches for plane stress as well as plane strain [15].

To this stage we have discussed the solution of the governing differential equations. This is done either directly for a few simple situations or by using a variety of analytical techniques or by numerical solution of the equations in finite-difference forms. However, there is an alternative and extremely useful approach to obtaining solutions in continuum mechanics which is based on the use of variational methods. A number of theorems have been developed for elasticity and plasticity on these bases which lead to minimization procedures or extremum principles. There is an extensive literature with the results often being presented in somewhat different ways by different authors. Here we will give only a qualitative discussion of some of the approaches that have been useful in fracture studies.

The starting point is the "principle of virtual work." We imagine a solid body to be given infinitesimal displacement $\delta u, \delta v, \delta w$ in the x, y, and z directions which satisfy compatibility and any displacement boundary conditions. These are referred to as "virtual displacements" since they are not necessarily those existing in the body but are invoked to induce a variation in work about the current stress and displacement fields. The "virtual work" done by the external forces is merely the summation of the work done by each force as it undergoes its virtual displacement. We assume also a state of stress which is only required to satisfy equilibrium and any force boundary conditions. Then the volume integral of the stresses times the corresponding strain increments obtained from the virtual displacements (i.e., the volume integral of δW_1 of Eq. (2.10a)) may be shown to equal the virtual work of the external forces. This derivation applies for any type of material behavior and for small strains. For elastic behavior, the virtual work done by the external forces equals the change δU_e in the elastic strain energy U_e. In this case δU_e can be recovered on unloading and can be regarded as the change in the potential energy of deformation [16]. Thus, for elastic behavior, the theorem of virtual work may be written as

$$\delta\left(U_e - \int_F \left(p_x u + p_y v + p_z w\right)\mathrm{d}A\right) = 0 \qquad (2.32a)$$

where p_x, p_y, p_z are surface forces per unit area, $\mathrm{d}A$ is the differential element of surface area, and the integral has taken over the surface area (F) upon which force boundary conditions are prescribed. The integral may be interpreted as the potential energy of the surface forces so the preceding equation becomes

$$\delta(\text{PE}) = 0 \tag{2.32b}$$

where PE is the total potential energy of the system in the absence of body forces. With the additional postulate of a "stable" material which includes Hooke's law but excludes stress–strain curves with negative slopes, it may be shown that the stationary condition on the potential energy is a minimum. Hence, for an elastic system, by giving a body in equilibrium virtual displacements, it is shown that the equilibrium state is one of "minimum potential energy." This potential energy is the sum of the elastic strain energy and that of the external loads. In Chap. 4 we will see that this concept forms the basis of the first analytical study of fracture.

A corresponding result may be derived for the "complementary energy" (CE) of the system by considering virtual increments in the stresses, which satisfy equilibrium and the boundary conditions leading to

$$\delta\left(U_c - \int_D (p_x u + p_y v + p_z w)\, dA \right) = 0 \tag{2.33a}$$

$$\text{or } \delta(\text{CE}) = 0 \tag{2.33b}$$

where the surface integral has now taken over the region (D) in which displacements are prescribed. In calculating the variation of the PE, the quantities p_x, p_y, p_z were held constant. Now in calculating the variation of the CE, the displacements u, v, and w are held constant. Again for a stable elastic material, it may be shown that at equilibrium the CE is not only stationary but is a minimum. Since the principles of minimum potential and complementary energy apply to nonlinear elasticity, from the analogy between incompressible, nonlinear elasticity and deformation theory of plasticity, the principles are sometimes invoked in dealing with the deformation theory of plasticity. However, this approach is unrealistic if unloading is involved.

Equations (2.32a) and (2.32b) may be shown [17] to lead to the extremely useful "Theorem of Castigliano." One statement of this is "knowing the complementary strain energy U_c of an elastic body in terms of the 'statically independent' loads $Q_1, Q_2, \cdots Q_i, \cdots Q_n$ applied to it," and we may calculate the displacement x_i under load Q_i from

$$x_i = \frac{\partial U_c}{\partial Q_i} \tag{2.34a}$$

For linear elastic behavior, $U_e = U_c$, and this may be written as

$$x_i = \frac{\partial U_e}{\partial Q_i} \tag{2.34b}$$

which is the form in which it was originally derived by Castigliano. The "loads" may be moments or torques in which case the corresponding displacements are angles of rotation. The statement that the loads must be "statically independent" requires that we apply force and moment equilibrium considerations to relate the various loads or moments before applying Castigliano's Theorem. Thus, we are dealing with forces which can be varied independently without altering the static equilibrium of the part.

From Eqs. (2.31a) and (2.31b), a counterpart of Castigliano's Theorem may be derived as

$$Q_i = \frac{\partial U_e}{\partial x_i} \tag{2.35a}$$

This holds for nonlinear or linear elasticity. For the linear elastic case with $U_e = U_c$, it may be written as

$$Q_i = \frac{\partial U_c}{\partial x_i} \tag{2.35b}$$

Another useful result is obtained by noting that if in Eq. (2.33a) the variations in stresses do not influence the boundary forces, then the second integral disappears and

$$\delta(U_c) = 0$$

For linear elastic behavior, $U_c = U_e$, and we have the statement that "For prescribed boundary conditions the stress distribution in a linear elastic body will minimize the strain energy." This result, sometimes referred to as the "principle of least work," finds wide application in stress analysis.

More general extremum principles can in fact be developed from Eqs. (2.32a), (2.32b), (2.33a), and (2.33b) for elastic material [3]. It may be shown that for all strain distributions which satisfy compatibility and displacement boundary conditions, the true distribution minimizes the potential energy and at the same time leads to stresses satisfying equilibrium. Similarly for all stress distributions satisfying equilibrium and force boundary conditions, the true distribution minimizes the complementary energy and leads to strains which satisfy compatibility. This opens up the possibility of numerical minimization techniques to provide solutions. The familiar "finite element" method is the most prominent example of such an approach. It has been described in detail in many books so we mention only that it involves splitting the body into small, usually compatible, elements whose displacements or stresses are adjusted on their boundaries at so-called nodes. The element equations are obtained upon application of a minimum or stationary principle governing stresses and/or displacements. The potential energy of the body, for example, is the sum of the element potential energies and becomes a function of the nodal values only. To find displacements which minimize the potential energy of the system within the limits of the finite element discretization, the relationship $\partial(\text{PE})/\partial u_j = 0$ is satisfied for each unconstrained nodal displacement u_j. The displacement u_j, which minimizes the PE, is used to calculate the strains, and finally the stresses are determined from the constitutive equation. Finite element methods have been used extensively to obtain elastic and elastic–plastic solutions for notched and cracked parts. A great deal of ingenuity has gone into circumventing the difficulty posed by the large stress or strain gradients at the crack tip. In particular, special "finite elements" have been devised to describe conditions at the crack tip. Several reviews [18–20] have been given of the different approaches which can be followed with finite elements to study fracture problems. We mention only that the "obvious" approach of taking very small elements to handle the large gradients in stress or strain at a crack tip has been shown to be very inefficient.

Variational methods have also been applied to problems involving plastic deformation. Before discussing these, it may be helpful to discuss first the concept of the "limit load."

As loads are increased in a part which is subjected to nonuniform stresses, an increasing amount of material will yield and undergo plastic deformation. In an elastic–perfectly plastic (non-strain-hardening) material, when the plastically deformed region is no longer constrained by adjacent material which has not yielded, the deformation becomes unbounded and the structure is said to collapse. The calculation of this limit or collapse load is an important problem in both structural design and metal forming. It is also of importance in the analysis of service failures for in many cases fracture follows rather than precedes the attainment of the limit load. While limit loads can be calculated readily for simple cases such as bending of beams or twisting of shafts, in other cases a complete solution can only be obtained by numerical means. Fortunately, variational methods have enabled theorems to be developed which place lower and upper bounds on the limit load without requiring a complete solution for stresses and displacements. These theorems are usually presented for elastically rigid–perfectly plastic material although an extension to the more general case of elastic–perfectly plastic is possible. The derivation of the theorems is again based on "virtual work" combined now with the "principle of maximum plastic resistance" discussed in Sect. 2.7. The mathematical derivation [21] of the theorems is clear-cut, but their description in words by different authors varies considerably. In essence, the theorems tell us

Lower bound: "If a limit load is calculated from assumed stresses which satisfy equilibrium and force boundary conditions and if these stresses are within or on the yield surface, it will be less than or equal to the true limit load." The state of stress need not be the exact one and the associated strains need not satisfy compatibility.

Upper bound: "If we assume a state of deformation which satisfies compatibility and displacement boundary conditions, then the load calculated by equating the work done by external loads to the work done internally, assuming that yield has occurred, will be greater than or equal to the true limit load." The deformation pattern assumed need not be the correct one, and associated stresses need not satisfy equilibrium.

Since the two bounds bracket the true limit load, one might attempt to determine the limit load by minimizing the upper bound and maximizing the lower bound. In practice this may involve so much effort that a numerical solution based on a more realistic stress–strain curve may be preferable.

It is intriguing that equivalent theorems may be stated for the case of a masonry or stonework structure which is infinitely strong in compression but has no tensile strength [22]. It may be shown that "If a line of thrust in a stone structure can be found which is in equilibrium with the applied loads and body forces, and lies wholly within the stonework, then the structure is safe." This need not be the actual line of thrust. Similarly there is an upper-bound theorem for stonework. "If enough hinges form to reduce the structure to a mechanism, by equating the work done by the internal forces, an upper bound is obtained." The postulated collapse mechanism need not be the actual one. We will return to these theorems at the end of Chap. 8 when we discuss the compressive behavior of brittle solids.

Our final topic in this section is the plane-strain "slip-line" approach to solutions for a rigid–perfectly plastic material. Although this approach preceded the lower- and upper-bound theorems, it is useful to compare it to them since they deal with the same material behavior. In many problems involving cracks or in metal working, the case of plane strain arises in which the strain in the direction of the intermediate principal stress σ_2 is zero. For incompressible material, $\sigma_2 = (1/2)(\sigma_1 + \sigma_3)$, and the state of stress $\sigma_1, \sigma_2, \sigma_3$ is equivalent to $(\sigma_1 - \sigma_3)/2, 0, (\sigma_3 - \sigma_1)/2$ plus a hydrostatic stress $\sigma_2 = (1/2)(\sigma_1 + \sigma_3)$. Thus, using the maximum shear stress yield criterion, yield and subsequent plastic deformation occur whenever $(\sigma_1 - \sigma_3)/2$ equals the yield strength κ in shear.

If instead of using a Cartesian coordinate system, one chooses a curvilinear coordinate system with the axes in the directions of the maximum shear stresses, then the equilibrium equations assume an extremely simple form. The coordinate system, formed by the maximum shear stress directions, is referred to as a slip-line field, and the two sets of orthogonal directions or slip lines are traditionally referred to as α and β lines. The maximum shear stress has the constant value κ along these lines, while the hydrostatic stresses change between two points on an α or β line by $\pm 2\kappa\phi$, respectively, where ϕ is the angle turned through by a line between the two points. Without going into details, in essence, the solution of the equilibrium equations and yield condition becomes a problem of drawing an orthogonal network subject to certain geometrical constraints. Since deformation along a slip line is pure shear, there are no normal displacements along any slip line. This imposes certain restrictions on the displacements. The construction of slip-line fields which satisfy equilibrium and compatibility in the yielded region and are surrounded by rigid material has been carried out for many configurations [23]. Some of these useful for fracture are given in the next chapter.

Slip-line solutions usually neglect the stress field in the rigid material, but the displacements are adequate for the computation of an upper bound. In this sense, a complete slip-line solution should provide an upper bound which may be the exact solution. On the other hand, if only the equilibrium equations are considered, the slip-line approach may be useful in establishing states of stress from which lower bounds can be obtained. It has been pointed out that slip-line solutions are not unique in that different deformation patterns can be found for the same loading. However, with the inclusion of a small amount of strain hardening, this lack of a unique solution disappears [24]. Although slip-line solutions are useful in certain fracture studies, we should caution that their predictions may depart significantly from more realistic numerical solutions which allow for strain hardening.

References

1. Jaeger JC. Elasticity, fracture and flow. London: Methuen; 1956.
2. Bridgeman PW. The stress distribution at the neck of a tensile specimen. Trans Am Soc Metals. 1944;32:553–74. See also his book: "Large Plastic Flow and Fracture", McGraw-Hill Book Co., New York 1952.
3. Hill R. The mathematical theory of plasticity. London: Oxford University Press; 1950.
4. Prager W, Hodge PG. Theory of perfectly plastic solids. New York: Wiley; 1951.
5. Bishop JFW, Hill R. A theory for the plastic distortion of a polycrystalline aggregate under combined stress. Philos Mag. 1951;42:414–27. 1298–1307.
6. McClintock FA, Argon AS. Mechanical behavior of materials. Reading, MA: Addison-Wesley Publishing Co.; 1966.
7. Finnie I, Heller WR. Creep of engineering materials. New York: McGraw-Hill Book Co., Inc.; 1959.
8. Timoshenko S, Goodier JN. Theory of elasticity. 3rd ed. New York: McGraw-Hill Book Co.; 1970.
9. Muskhelishvili NI. Some basic problems of the mathematical theory of elasticity. Groningen, The Netherlands: Noordhoff Ltd; 1953 (published originally in Russian in 1933).
10. Westergaard HM. Bearing pressure and cracks. Trans Am Soc Mech Eng J Appl Mech. 1939;6:A49–53.
11. Willians ML. On the stress distribution at the base of a stationary crack. Trans ASME J Appl Mech. 1957;24:109–14.
12. Sih GC, editor. Mechanics of fracture I—methods of analysis and solution of crack problems. Leyden, The Netherlands: Noordhoff Int. Pub; 1973.
13. Srawley JE, Gross B. Stress intensity factors for bend and compact specimens. Eng Frac Mech. 1972;4:587–9.
14. Southwell RV. Relaxation methods in theoretical physics. Oxford: Oxford University Press; 1946.
15. Allen DN d G, Southwell RV. Relaxation methods applied to engineering problems, XIV, Plastic straining in two-dimensional stress systems. Philos Trans R Soc Lond. 1950; A242:379–414.
16. Flugge CW. Handbook of engineering mechanics. New York: McGraw-Hill; 1962.
17. Langhaar HL. Energy methods in applied mechanics. New York: Wiley; 1962.
18. Becker EB, Dunham RS, Stern M. Some stress intensity calculations using finite elements. In: Pulmano V, Kabaila A, editors. Finite element methods in engineering. Sydney: Clarendon Press; 1974. p. 117–38.
19. ASME. Computational fracture mechanics. Special volume of papers at 2nd Nat. Congress on Pressure Vessels and Piping, San Francisco, 23–27 Jun 1975.

20. Gallagher RH. Survey and evaluation of the finite element method in linear fracture mechanics analysis. Proc. 1st Int. Conf on Structural Mech. in Reactor Technology, Vol. 6, Part L, Sept 1972, p. 637–648.
21. Johnson W, Mellor PB. Engineering plasticity. London: Van Nostrand Reinhold; 1973.
22. Kooharian A. Limit analysis of voussoir (segmental) and concrete arches. J Am Concrete Inst. 1953;49:317–28. discussion 328-1–328-4.
23. Johnson W, Sowerby S, Haddow JB. Plane-strain slip line fields: theory and bibliography. New York: Americal Elsevier Publishing Co., Inc.; 1970.
24. McClintock FA. Plasticity aspects of fracture, Chap. 2. In: Liebowitz H, editor. Fracture, vol. 3. New York: Academic; 1971. p. 48–225.

Chapter 3
Some Solutions That Are Useful in Fracture Prediction

3.1 Introduction

Our primary objective is to use solutions that have been developed for elastic, elastic–plastic, and fully plastic behavior to predict fracture. The detailed development of these solutions and a discussion of techniques, analytical and numerical, used to obtain them are beyond the scope of this work. However, for completeness, some of the analytical techniques that have been used to obtain stress fields for sharp cracks and other configurations will be described. We consider first the elastic solutions for sharp cracks, circular holes, and elliptical holes in flat plates. Also, the elastic solutions for cylindrical inclusions for spherical holes and inclusions will be summarized. Then, approximate and more exact solutions, based on elastic–plastic behavior for sharp and blunted cracks, will be described. Finally, solutions for notched members will be presented, both for the initial elastic stress distribution and for the case in which yielding has occurred over the entire cross section. A few solutions have been omitted from this chapter because it seems more appropriate to treat them at a later stage. These are the Dugdale model for the plastic zone ahead of a crack for plane stress (Chap. 5), the solutions for stresses and strains (in terms) of the J integral (Chap. 6), and the solutions for void growth in ductile fracture (Chap. 7).

Much of our subsequent treatment of fracture will be based on the elastic solutions for the stresses near the tip of the crack shown in Fig. 3.1. Since these solutions are used by many people who are not concerned with the details of their derivation, they are summarized in Sect. 3.2 before outlining their derivation in Sect. 3.3. The coordinate axes in Fig. 3.1 are located at the tip of the crack, shown shaded, which extends through the thickness of the plate. The crack lies along the negative x axis, the y axis is normal to the plane of the crack, and the z axis coincides with the leading edge of the crack. *Mode I* or *direct opening* loading shown in Fig. 3.2 applies to the majority of fracture problems that will be discussed, and the stresses near the crack tip will be defined for this case. **Mode II or forward shear** loading occurs less frequently, but combinations of Mode I and Mode II arise

© Springer Science+Business Media New York 2016
C.K.H. Dharan et al., *Finnie's Notes on Fracture Mechanics*,
DOI 10.1007/978-1-4939-2477-6_3

Fig. 3.1 Crack in c plate

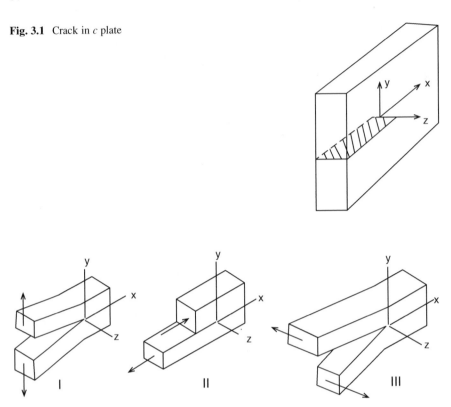

Fig. 3.2 The basic modes of crack surface displacements

when the loading axis is inclined to the plane of the crack. **Mode III, described as antiplane strain or sideways shear**, is by far the easiest of the modes to analyze, particularly if plastic as well as elastic deformation is involved. A practical example of **Mode III loading is the circumferentially notched round shaft loaded in torsion**. As pointed out in Chap. 1, the slant fracture, shown in Fig. 1.3, which results from the initial Mode I loading of a thin sheet is a combination of Mode I and III loading. As with all solutions for linear elastic behavior, the stresses, strains, or displacements for Modes I, II, and III loading may be superimposed unless buckling occurs or if the boundary conditions change with load.

3.2 State of Stress Near the Tip of the Crack

If only the region near the crack tip is considered, remarkably simple expressions are obtained for the stress components. These are shown in Fig. 3.3 when the element upon which the stresses act is located by the coordinates, r, θ measured in the x–y plane. For all three modes, the stress components that are not zero have the form $\frac{K}{(2\pi r)^{1/2}} f(\theta)$ where K, the stress intensity factor, is written as K_{I}, K_{II}, or K_{III}

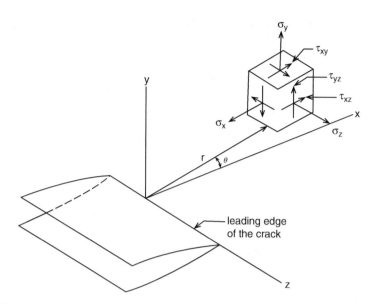

Fig. 3.3 Coordinates measured from the leading edge of a crack and the stress components in the crack tip stress field

depending on the mode of loading involved. Only $f(\theta)$, the function of angle, differs for the stress components for a given mode of loading. Stress intensity factors, which have dimensions $MN/m^{3/2} = MPa\text{-}m^{1/2}$ (or $psi\text{-}in^{1/2}$), should not be confused with the stress concentration factor, usually denoted by K_t, which is dimensionless. Stress intensity factors, as we will see, have been obtained for a great variety of cracked configurations. In general, the numerical value of the stress intensity factor depends on the dimensions of the part, the size of the crack, and the applied loads or displacements. Thus, for the large variety of cracked configurations which involve Mode I loading, the stress distributions near the crack tip differ only by a "scale factor" which is determined by K_I. The same statement applies to Mode II or Mode III loading with K_I replaced by K_{II} or K_{III}. Similarly, the displacement components near the crack tip have the forms $\frac{K}{G}\left(\frac{2r}{\pi}\right)^{1/2}g(\theta)$ where G is the shear modulus and $g(\theta)$ follows the description given for $f(\theta)$.

For Mode I and Mode II loading applied to "thin" plates, the condition is one of the plane stresses in which $\sigma_z = 0$ and $\varepsilon_z = -(\nu/E)\left(\sigma_x + \sigma_y\right)$. However, in the other extreme, with thick enough plates, the large bulk of material under very low stresses which surrounds the crack tip region will prevent contraction of expansions in the z direction. This is a special case of plane strain in which $\varepsilon_z = 0$ and hence $\sigma_z = \nu\left(\sigma_x + \sigma_y\right)$ for elastic behavior. The distinction between plane stress and plane strain is an important one in fracture, and at a later stage, the actual values of plate thickness corresponding to these conditions will be discussed.

In the solutions that follow, the displacements u, v in the x, y directions are given for plane strain. To obtain the displacements for plane stress, Poisson's ratio is replaced by $\nu/(\nu + 1)$. This is equivalent to replacing ν in solutions for plane stress by $\nu/(1 - \nu)$ to obtain solutions for plane strain. The displacement w in the z direction is zero for plane stress for Modes I and II. It may be obtained from ε_z and the plate thickness for plane stress:

Mode I: (Plane strain)

$$\sigma_x = \frac{K_1}{(2\pi r)^{1/2}} \cos\frac{\theta}{2}\left[1 - \sin\frac{\theta}{2}\sin\frac{3\theta}{2}\right]$$

$$\sigma_y = \frac{K_1}{(2\pi r)^{1/2}} \cos\frac{\theta}{2}\left[1 + \sin\frac{\theta}{2}\sin\frac{3\theta}{2}\right]$$

$$\tau_{xy} = \frac{K_1}{(2\pi r)^{1/2}} \sin\frac{\theta}{2}\cos\frac{\theta}{2}\cos\frac{3\theta}{2} \tag{3.1}$$

$$\tau_{xz} = \tau_{yz} = 0$$

$$u = \frac{K_1}{G}\left(\frac{r}{2\pi}\right)^{1/2} \cos\frac{\theta}{2}\left[1 - 2\nu + \sin^2\frac{\theta}{2}\right]$$

$$v = \frac{K_1}{G}\left(\frac{r}{2\pi}\right)^{1/2} \sin\frac{\theta}{2}\left[2 - 2\nu - \cos^2\frac{\theta}{2}\right]$$

$$\boxed{\begin{aligned}
\sigma_{1,2} &= \frac{\sigma_x + \sigma_y}{2} \pm \sqrt{\left(\frac{\sigma_x - \sigma_y}{2}\right)^2 + \tau_{xy}^2} \\[4pt]
\sigma_1 &= \frac{K_1}{(2\pi r)^{1/2}} \cos\frac{\theta}{2}\left[1 + \sin\frac{\theta}{2}\right] \\[4pt]
\sigma_2 &= \frac{K_1}{(2\pi r)^{1/2}} \cos\frac{\theta}{2}\left[1 - \sin\frac{\theta}{2}\right] \\[4pt]
\sigma_3 &= 0 \text{ (plane stress) or } \sigma_3 = \frac{2\nu K_1}{(2\pi r)^{1/2}} \cos\frac{\theta}{2} \text{ (plane strain)}
\end{aligned}}$$

Mode II: (Plane strain)

$$\sigma_x = \frac{K_{II}}{(2\pi r)^{1/2}} \sin\frac{\theta}{2}\left[2 + \cos\frac{\theta}{2}\cos\frac{3\theta}{2}\right]$$

$$\sigma_y = \frac{K_{II}}{(2\pi r)^{1/2}} \sin\frac{\theta}{2}\left[\cos\frac{\theta}{2}\cos\frac{3\theta}{2}\right]$$

$$\tau_{xy} = \frac{K_{II}}{(2\pi r)^{1/2}} \cos\frac{\theta}{2}\left[1 - \sin\frac{\theta}{2}\sin\frac{3\theta}{2}\right] \tag{3.2}$$

$$\tau_{xz} = \tau_{yz} = 0$$

$$u = \frac{K_{II}}{G}\left(\frac{r}{2\pi}\right)^{1/2} \sin\frac{\theta}{2}\left[2 - 2\nu + \cos^2\frac{\theta}{2}\right]$$

$$v = \frac{K_{II}}{G}\left(\frac{r}{2\pi}\right)^{1/2} \cos\frac{\theta}{2}\left[-1 + 2\nu + \sin^2\frac{\theta}{2}\right]$$

Mode III: (Plane strain & Plane stress)

$$\tau_{xz} = \frac{K_{III}}{(2\pi r)^{1/2}} \sin\frac{\theta}{2}$$

$$\tau_{yz} = \frac{K_{III}}{(2\pi r)^{1/2}} \cos\frac{\theta}{2}$$

$$\sigma_x = \sigma_y = \sigma_z = \tau_{xy} = 0 \tag{3.3}$$

$$w = \frac{K_{III}}{G}\left(\frac{2r}{\pi}\right)^{1/2} \sin\frac{\theta}{2}$$

$$u = v = 0$$

In all cases, the solutions apply for $-180° \leq \theta \leq 180°$.

A caution in connection with Eqs. (3.1)–(3.3) is that some authors absorb the constant π or 2π into the value of the stress intensity factor. For example, if stresses are written as $\frac{k}{(2r)^{1/2}}f(\theta)$, then $k = \frac{K}{\pi^{1/2}}$. The notation of Eqs. (3.1)–(3.3) is the one most commonly employed and is used by American Society for Testing and Materials (ASTM).

The solutions that have been summarized for Modes I, II, and III can only be valid near the crack tip since they tend to zero when r tends to infinity. In general, the complete solution for a stress component can be expressed by a series in which the successive terms involve $r^{-1/2}$, r^0, $r^{1/2}$, r, $r^{3/2}$, etc. We have been discussing only the first of these, the "singularity" term which dominates as r tends to zero. The region over which the first term provides a close approximation to the exact solution and the role of the other terms in the series expansion are important topics for later discussion.

3.3   Determination of the Stresses for Mode I and Mode II Loading

Closed form analytical solutions for these modes of loading are generally limited to infinite plates with practical specimen geometries requiring numerical solution. However, the analytical solutions are useful in indicating the details of the stress field near the crack tip and as a starting point for approximate or numerical solutions.

The most general approach, which is associated with Muskhelishvili [1], expresses solutions of the biharmonic equations in terms of two functions of the complex variable. However, for cases in which the crack occupies a section of the x axis, a simpler approach due to Westergaard [2], which involves only one function of the complex variable, allows useful solutions to be obtained. Before launching into Westergaard's derivation which is adequate for our purposes, it may be worthwhile to review some aspects of complex variables. First, a function of a complex variable, which is single valued throughout a region, is said to be analytic throughout that region if it has a derivative at every point of that region.

The derivative of a function $f(z)$ of the complex variable $z = x + iy$ is defined by the same formula as for a real variable:

$$f'(z) = \frac{d}{dx}f(z) = \lim_{\Delta z \to 0} \frac{f(z + \Delta z) - f(z)}{\Delta z} = \lim_{\Delta z \to 0} \frac{\Delta f}{\Delta z}$$

However, the existence of the derivative of a complex function at a point requires that $\frac{\Delta f}{\Delta z} = \frac{\Delta f}{\Delta x + i \Delta y}$ has the same limit no matter how Δx and Δy approach zero. If $f(z) = f_1(x, y) + if_2(x, y)$ where f_1 and f_2 are real functions of x and y, the derivative $f'(z)$ may be computed by letting first Δy and then Δx approach zero. This leads to

$$f'(z) = \frac{\partial f_1}{\partial x} + i\frac{\partial f_2}{\partial x} \qquad (3.4a)$$

Alternatively, letting Δx approach zero first and then Δy leads to

$$f'(z) = -i\frac{\partial f_2}{\partial y} + \frac{\partial f_1}{\partial y} \qquad (3.4b)$$

The equivalence of these two expressions for $f'(z)$ requires

$$\frac{\partial f_1}{\partial x} = \frac{\partial f_2}{\partial y} \qquad \frac{\partial f_1}{\partial y} = \frac{\partial f_2}{\partial x}$$

These are known as the Cauchy–Riemann equations. It can be shown that if $f(z)$ is analytic throughout a region, not only the first partial derivatives of f_1 and f_2, but also all those of higher orders exist and are analytic throughout the region. Thus, the first of the Cauchy–Riemann equations can be differentiated with respect to x and the second with respect to y and the equations subtracted to obtain

$$\frac{\partial^2 f_1}{\partial x^2} + \frac{\partial^2 f_1}{\partial y^2} = 0, \qquad \nabla^2 f_1 = 0$$

Similarly, differentiating the first with respect to y, the second with respect to x, and adding lead to

$$\frac{\partial^2 f_2}{\partial x^2} + \frac{\partial^2 f_2}{\partial y^2} = 0, \qquad \nabla^2 f_2 = 0$$

That is, both f_1 and f_2 satisfy Laplace's equation. A solution of this equation with continuous second partial derivatives is called a harmonic function. Such a function is an acceptable stress function, for if $\nabla^2 f_1 = 0$ everywhere in a region, then $\nabla^4 f_1 = \nabla^2(\nabla^2 f_1) = 0$. However, one can be less restrictive in choosing stress functions, for if ϕ_1, ϕ_2, and ϕ_2 are all harmonic functions, then $\psi = \phi_1 + x\phi_2 + y\phi_3$ or $\psi = (x^2 + y^2)\phi_1$ will be acceptable stress functions.

Westergaard chose the notation $\overline{\overline{Z}}(z)$ for an analytic function with $\overline{Z}(z)$, $Z(z)$, and $Z'(z)$ denoting successive derivatives (i.e., $\frac{d\overline{\overline{Z}}}{dz} = \overline{Z}$, $\frac{d\overline{Z}}{dz} = Z$, $\frac{dZ}{dz} = Z'$). Since the real and imaginary parts of $\overline{\overline{Z}}$ and its derivatives are harmonic functions, they may be used to form a stress function. Equations (3.4a) and (3.4b) may be used to differentiate these functions. For example, using the notation

$$f_1 = \mathrm{Re}\overline{\overline{Z}}, f_2 = \mathrm{Im}\overline{\overline{Z}}$$

$$\mathrm{Re}\overline{Z} = \frac{\partial \mathrm{Re}\overline{\overline{Z}}}{\partial x} = \frac{\partial \mathrm{Im}\overline{\overline{Z}}}{\partial y}$$

$$\mathrm{Im}\overline{Z} = \frac{\partial \mathrm{Im}\overline{\overline{Z}}}{\partial x} = \frac{\partial \mathrm{Re}\overline{\overline{Z}}}{\partial y}$$

$$\overline{Z} = \mathrm{Re}\overline{Z} + i\mathrm{Im}\overline{Z} = \frac{d\overline{\overline{Z}}}{dZ}\left(= \frac{d}{dZ}\left\{\mathrm{Re}\overline{\overline{Z}} + i\mathrm{Im}\overline{\overline{Z}}\right\}\right)$$
$$= \frac{\partial \mathrm{Re}\overline{\overline{Z}}}{\partial x} + i\frac{\partial \mathrm{Im}\overline{\overline{Z}}}{\partial x} = -i\frac{\partial \mathrm{Re}\overline{\overline{Z}}}{\partial y} + \frac{\partial \mathrm{Im}\overline{\overline{Z}}}{\partial y}$$
$$(dZ = dx + idy)$$

For Mode I loading, Westergaard chose as his stress function

$$\psi_1 : \mathrm{Re}\overline{\overline{Z}} + y\mathrm{Im}\overline{Z}$$

which leads to stresses[1], recalling Eqs. (2.28) and (2.29):

$$\sigma_x = \frac{\partial^2 \psi_1}{\partial y^2} = \mathrm{Re}Z - y\mathrm{Im}Z'$$

$$\sigma_y = \frac{\partial^2 \psi_1}{\partial x^2} = \mathrm{Re}Z - y\mathrm{Im}Z' \qquad (3.5)$$

$$\tau_{xy} = \frac{\partial^2 \psi_1}{\partial x \partial y} = -\mathrm{Re}Z'$$

For this stress function for $y = 0$ (i.e., along the x axis), $\tau_{xy} = 0$, $\sigma_x = \sigma_y$. The displacements u, v in the x, y directions for plane strain may be obtained from the expressions for the strains in Sect. 2.9. Starting with

$$2G\varepsilon_x^e = \sigma_x(1 - \nu) - \nu\sigma_y$$

---

[1] It was pointed out by Sih, and later by Eftis and Liebowitz, that Westergaard's original formulation is in error. The expressions for σ_x and σ_y should include $+A$ and $-A$, respectively. A vanishes for crack line loading and is zero for biaxial tension at infinity.

and integrating after putting

$$\varepsilon_x^e = \frac{\partial u}{\partial x} \text{ leads to}$$

$$2Gu = \int \sigma_x(1 - \nu)\partial x - \nu \int \sigma_y \partial x$$

Substituting for the stresses from Eq. (3.5) and making use of the expressions given earlier for the derivatives of real and imaginary parts of complex function yield

$$2Gu = (1 - 2\nu)\mathrm{Re}\overline{Z} - y\mathrm{Im}Z \qquad (3.6)$$

Similarly, $2Gv = 2(1 - \nu)\mathrm{Im}\overline{Z} - y\mathrm{Re}Z$
    In particular, for $y = 0$,

$$v = \frac{(1 - \nu)}{G}\mathrm{Im}\overline{Z} = \frac{2(1 - \nu^2)}{E}\mathrm{Im}\overline{Z}$$

At this point, any analytic function (i.e., single valued with derivatives) $Z(z)$ will give stresses which satisfy the equations of elasticity for a specific set of boundary conditions. Before looking at a complete solution, it is of interest to examine the stress distribution close to the crack tip to confirm the expression presented in Eq. (3.1). For the crack in Mode I loading of Fig. 3.2, the function $Z(z)$ should be analytic except for a branch cut along the negative x axis. This is given by a function of the form $Z(z) = \frac{f(z)}{z^{1/2}}$ provided that the imaginary part of $f(z)$ is zero along the negative x axis so that $y = 0$ on the crack faces. From Eq. (3.5), it is seen that $\tau_{xy} = 0$ along the crack faces as $y = 0$. In fact, since the stresses tend to infinity as the crack tip is approached, the function $f(z)$ must approach a real constant as $|Z| \to 0$. Taking the constant as $f(z) = \frac{K_1}{(2\pi r)^{1/2}}$ to be consistent with the expressions for the stresses given earlier leads to

$$Z(z) \underset{|Z| \to 0}{=} \frac{K_1}{(2\pi z)^{1/2}} = \frac{f(z)}{z^{1/2}} \qquad (3.7)$$

Making the substitution $z = re^{i\theta}$,

$$Z(z) = \frac{K_1}{(2\pi r)^{1/2}}\left(\cos\frac{\theta}{2} - i\sin\frac{\theta}{2}\right)$$

$$Z'(z) = -\frac{1}{2}\frac{K_1}{(2\pi)^{1/2}z^{3/2}} = -\frac{1}{2}\frac{K_1}{(2\pi r)^{1/2}r}\left(\cos\frac{2\pi}{2} - i\sin\frac{3\theta}{2}\right)$$

Using these expressions and $y = r \sin\theta$ with Eq. (3.5) leads to the stresses near the tip of the crack given in Eq. (3.1).

A case studied by Westergaard, which we shall use extensively in later work, is that of a crack along the x axis from $x = -a$ to $x = +a$ in an infinite plate loaded by biaxial tensions σ on its boundary. He showed that the solution is obtained from a stress function

$$Z(z) = \frac{\sigma z}{(z^2 - a^2)^{1/2}} = \frac{\sigma}{(1 - a^2/z^2)^{1/2}} \tag{3.8}$$

The boundary conditions at infinity can be seen to be satisfied since

$$Z'(z) = -\frac{-\sigma a^2}{(z^2 - a^2)^{3/2}} \to 0 \text{ and } \mathrm{Re}Z(z) \to \sigma \text{ as } z \to \infty.$$

Since $\tau_{xy} = 0$ for $y = 0$, the only remaining boundary conditions to be confirmed are that $\sigma_y = 0$ is zero on the faces of the crack. This requires that $\mathrm{Re}Z$ be zero in this region which follows from Eq. (3.8). The amount by which the initially closed crack opens under load is given by twice the v displacement for $y = 0$ and $-a < x < a$ from Eq. (3.6). Since $\overline{Z}(z) = \sigma(z^2 - a^2)^{1/2} = i\sigma(a^2 - x^2)^{1/2}$ for $y = 0$, the crack opening is

$$2v(x) = \frac{4\sigma(1 - \nu^2)}{E}(a^2 - x^2)^{1/2} \quad -a < x < a \tag{3.9}$$

Thus, the line crack in an unloaded condition opens under load t_0 from an ellipse with a maximum opening at $x = 0$ of $\frac{2(1-\nu^2)\sigma a}{E}$ for plane strain.

The value of the stress intensity factor corresponding to the stress function $Z(z)$ in Eq. (3.8) can be found by choosing a new coordinate system located at the crack tip and comparing the value of $Z(z)$ as $|Z| \to 0$ with Eq. (3.4b). Letting $z - a = z*$ where $z*$ corresponds to coordinates measured from the crack tip as in Eq. (3.4b), Eq. (3.8) becomes

$$Z = \frac{\sigma(z* + a)}{[z*(z* + 2a)]^{1/2}} \simeq \sigma\left(\frac{a}{2z*}\right)^{1/2} \quad \text{as} \quad |z*| \to 0$$

Comparing with Eq. (3.7) shows that $K_I = \sigma(\pi a)^{1/2}$. Alternatively, an expression for one of the stress components could be obtained for the region near the crack tip and compared with Eq. (3.1) to determine K_I.

Superposition of a uniform compressive stress $\sigma_x = \sigma$ (parallel to the plane of the crack) does not change the singularity term. Thus, the stress intensity factor $K_I = \sigma$ $(\pi a)^{1/2}$ also applies to an infinite plate subjected to a uniform uniaxial tension σ normal to the plane of the crack with length 2a.

The procedure for a Mode II crack is very similar. Choosing a stress function

$$\psi_{\mathrm{II}} = -y\mathrm{Re}\overline{Z}$$

the stresses are found to be

$$\sigma_x = 2\mathrm{Im}Z + y\mathrm{Re}Z'$$

$$\sigma_y = -y\mathrm{Re}Z'$$

Again, in the neighborhood of the crack tip shown in Fig. 3.2, it is argued that $Z(z) \underset{|Z|\to 0}{=} \frac{K_{\mathrm{II}}}{(2\pi z)^{1/2}}$ which leads to the expression given earlier for the crack tip stress field. Since the two stress functions ψ_{I} and ψ_{II} can be added to form a permissible stress function $\psi_{\mathrm{I}} + \psi_{\mathrm{II}}$, solutions for Mode I and Mode II loading on a given cracked part may be superimposed.

Another approach to the derivation of stresses for Mode I and Mode II cracks, due to Willians [3], makes use of the equation of elasticity in conventional form and leads to results which are more general than Westergaard's.

We start with the equations of two-dimensional elasticity in polar coordinates r and θ. The biharmonic $\nabla^4\psi = 0$ becomes

$$\left(\frac{\partial^2}{\partial r^2} + \frac{1}{r}\frac{\partial}{\partial r} + \frac{1}{r^2}\frac{\partial^2}{\partial\theta^2}\right)\left(\frac{\partial^2\psi}{\partial r^2} + \frac{1}{r}\frac{\partial\psi}{\partial r} + \frac{1}{r^2}\frac{\partial^2\psi}{\partial\theta^2}\right) = 0 \qquad (3.10)$$

and the stresses are

$$\sigma_r = \left(\frac{1}{r^2}\right)\left(\frac{\partial^2}{\partial\theta^2}\right) + \left(\frac{1}{r}\right)\frac{\partial\psi}{\partial r}$$

$$\sigma_\theta = \frac{\partial^2\psi}{\partial r^2}$$

$$\tau_{r\theta} = -\left(\frac{1}{r}\right)\left(\frac{\partial^2\psi}{\partial r\partial\theta}\right) + \left(\frac{1}{r^2}\right)\frac{\partial\psi}{\partial\theta}$$

Williams started by taking permissible stress functions for an elastic wedge of included angle α and then let $\alpha = 2\pi$ to obtain the cracked configuration of Fig. 3.4. For this situation, he showed that the stress function could be split into an even part ψ_e and an odd part ψ_0 with respect to θ:

Fig. 3.4 Stress distribution
at tip of crack using polar
coordinates

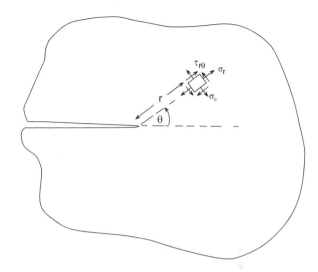

$$\psi_e = \sum_{i=1}^{n} \left\{ (-1)^{i-1} a_{2i-1} r^{i+1/2} \left[-\cos\left(i-\frac{3}{2}\right)\theta + \frac{2i-3}{2i+1} \cos\left(i+\frac{1}{2}\right)\theta \right] \right.$$
$$\left. + (-1)^{i} a_{2i} r^{i+1} [-\cos(i-1)\theta + \cos(i+1)\theta] \right\}$$

$$\psi_0 = \sum_{i=1}^{n} \left\{ (-1)^{i-1} b_{2i-1} r^{i+1/2} \left[\sin\left(1-\frac{3}{2}\right)\theta - \sin\left(i+\frac{1}{2}\right)\theta \right] \right.$$
$$\left. + (-1)^{i} b_{2i} r^{i+1} \left[-\sin(i-1)\theta + \frac{i-1}{i+1} \sin(i+1)\theta \right] \right\}$$

The constants a_i and b_i depend on the boundary conditions, and a common method of numerical solution (boundary collocation) is to select the n coefficients of a truncated series to give a "best fit" to the boundary conditions at a selected number $n/2$ of points. In examining the expressions for the stresses near the crack tip, it is seen that the leading terms, $a_1 r^{-1/2}$ and $b_1 r^{-1/2}$, dominate, and usually only the values of a_1 and b_1 are reported. Taking the first two terms in the expression for ψ, Williams obtained

$$\sigma_r = \frac{1}{4r^{1/2}} \left\{ a_1 \left[-5\cos\frac{\theta}{2} + \cos\frac{3\theta}{2} \right] + b_1 \left[-5\sin\frac{\theta}{2} + 3\sin\frac{3\theta}{2} \right] \right\} + 4a_2 \cos^2\theta$$

$$\sigma_\theta = \frac{1}{4r^{1/2}} \left\{ a_1 \left[-3\cos\frac{\theta}{2} - \cos\frac{3\theta}{2} \right] + b_1 \left[-3\sin\frac{\theta}{2} - 3\sin\frac{3\theta}{2} \right] \right\} + 4a_2 \sin^2\theta$$

$$\tau_{r\theta} = \frac{1}{4r^{1/2}} \left\{ a_1 \left[-\sin\frac{\theta}{2} - \sin\frac{3\theta}{2} \right] + b_1 \left[\cos\frac{\theta}{2} + 3\cos\frac{3\theta}{2} \right] \right\} + 4a_2 \sin 2\theta$$

$$(3.11)$$

The coefficient b_2 is found to be zero. The term involving a_2 will be referred to later in Chap. 4 in discussing crack stability. Superposition of a uniform stress σ_x in the x direction, as pointed out, will have no influence on the first, or singular, terms in the series expansions for the stresses. However, the coefficient a_2 of the second term will be changed to $a_2 + \sigma_x/4$ where σ_x is treated as positive when tensile. Apart from the term involving a_2, Eq. (3.11) is identical to Eqs. (3.1) and (3.2) if

$$a_1 = \frac{K_I}{(2\pi)^{1/2}} \quad \text{and} \quad b_1 = \frac{K_{II}}{(2\pi)^{1/2}}$$

3.4 Determination of the Stresses for a Mode III Crack

This case is considerably simpler than Modes I and II. Referring to Fig. 3.5, it is seen that $u = v = 0$, $w = w(x, y)$. Hence from the equations which relate strains to displacements,

$$\varepsilon_x = \varepsilon_y = \varepsilon_z = \gamma_{xy} = 0$$

$$\gamma_{xz} = \frac{\partial w}{\partial x}; \quad \gamma_{yz} = \frac{\partial w}{\partial y}$$

Hence,

$$\sigma_x = \sigma_y = \sigma_z = \tau_{xy} = 0$$
$$\tau_{xy} = G\gamma_{xz}; \quad \tau_{yz} = G\gamma_{yz}$$

The equilibrium equation is

$$\frac{\partial \tau_{xz}}{\partial x} + \frac{\partial \tau_{yz}}{\partial y} = 0$$

Fig. 3.5 Mode III loading. Directions of uvw displacements

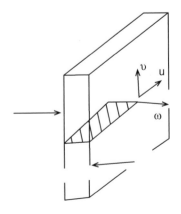

Fig. 3.6 Torsion
of a circumferentially
notched shaft. An example
of a Mode III crack

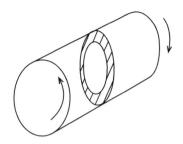

which in terms of the displacement, w is $\frac{\partial^2 w}{\partial x^2} + \frac{\partial^2 w}{\partial y^2} = 0$. That is, the z direction
displacement w must satisfy Laplace's equation. Choosing $w = (1/G)\mathrm{Im}\overline{Z}$ leads to
$\tau_{xz} = \mathrm{Im}Z, \tau_{yz} = \mathrm{Re}Z$. Arguing as before that $Z(z) = \dfrac{K_{\mathrm{III}}}{(2\pi r)^{1/2}}$ leads to the stresses and
$_{|Z|\to 0}$
displacements given in Eq. (3.3). An example of a situation in which Mode III
loading arises is shown in Fig. 3.6.

Apart from the simple case of the center cracked plate, the specific values of K_{I},
K_{II}, and K_{III}, which correspond to given loadings, crack lengths, and geometrical
shapes, have not been discussed. At least three extensive tabulations of stress
intensity factors are available in the literature [4–6], and a number of examples
will be given in Chap. 4.

3.5 Determination of the Stresses Around a Circular Hole

A solution which predates those for sharp cracks is for a plate with a small hole of
radius a through its thickness; the plate is loaded in its plane so that the state
of stress at large distances from the hole is uniaxial tension. This situation is not of
great importance in fracture, but it demonstrates some of the techniques of analysis
and some features of the stress field around a discontinuity in a simple manner.

Figure 3.7 shows the problem being considered and the coordinate system used
for the stresses. As in the William's solution for the sharp crack, polar coordinates
$(\sigma_\theta, \sigma_r, \tau_{r\theta})$ are the logical choice, and the biharmonic equation in the form of
Eq. (3.10) is used for the solution.

The stresses σ_θ and $\tau_{r\theta}$ at the surface of the hole, $r = a$, must be zero. Also, at
large distances from the hole, as $r \to \infty$, the state of stress must be uniaxial tension
of magnitude S; thus, σ_r, σ_θ, and $\tau_{r\theta}$ at infinity are known. These are the boundary
conditions of the problem.

It has been found [7] that a stress function of the form $\psi = f(r) + g(r)\cos 2\theta$
is suitable; $f(r)$ and $g(r)$ being functions of r are to be determined. They are
determined by substituting the preceding expression for ψ into the biharmonic

Fig. 3.7 Coordinate system

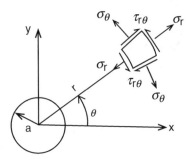

equation and solving the two ordinary differential equations which result; the constants of integration are evaluated from the boundary conditions. The final expressions for the stresses are

$$\sigma_r = \frac{S}{2}\left(1 - \frac{a^2}{r^2}\right) - \frac{S}{2}\left(1 + \frac{3a^4}{r^4} - \frac{4a^2}{r^2}\right)\cos 2\theta$$

$$\sigma_\theta = \frac{S}{2}\left(1 + \frac{a^2}{r^2}\right) + \frac{S}{2}\left(1 + \frac{3a^4}{r^4}\right)\cos 2\theta \qquad (3.12)$$

$$\tau_{r\theta} = \frac{S}{2}\left(1 - \frac{3a^4}{r^4} + \frac{2a^2}{r^2}\right)\sin 2\theta$$

Figure 3.8 shows the two components σ_x, $\sigma_y(\sigma_r, \sigma_\theta)$ along the x and y axes near the hole.

This solution is typical of those for stress fields near discontinuities and shows that the disturbance caused by the hole dies out very rapidly with increasing distance r. The stress concentration factor (maximum local stress $\div$ normal stress) is $\sigma_\theta(r = a, \theta = 0) \div \sigma = 3$. However, for $\theta = 0$ and a distance $r = 2a$, the value of σ_θ has dropped to 1.22 σ. For $\theta = \frac{\pi}{2}$ and $r = a$, the stress concentration factor is -1, and this local compressive stress can be relieved by buckling in thin sheets. Elastic solutions may be superimposed except where nonlinear behavior occurs, as in contact stress problems or buckling. Thus, if 3 and -1 are the maximum and minimum stress concentration factors at the hole for uniaxial tension, the value for biaxial tension as compression applied to a plate with a small hole is 2, while for pure shear, the stress concentration factors are 4 and -4.

The stresses of Eq. (3.12) apply for plane stress or plane strain. In thin sheets, $\sigma_z = 0$, while in thick enough sheets, at some distance from the free surface, the strain in the z direction at the hole must be equal to that away from the hole. The transverse stream $\varepsilon_z = \dfrac{\sigma - \nu(\sigma + \sigma_\theta)}{E}$ has a value $\dfrac{-\nu\sigma}{E}$ far from the hole. Thus, for plane strain conditions, taking σ_r and σ_θ from Eqs. (3.12) and equating the transverse strains near the hole and far from the hole:

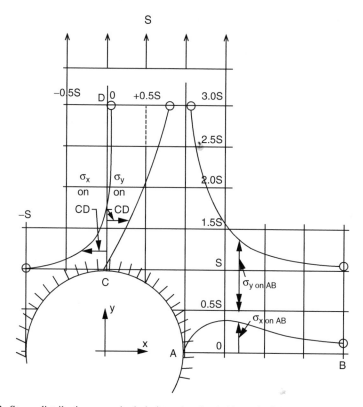

Fig. 3.8 Stress distribution around a hole in a plate loaded by uniaxial tension

$$\sigma_z = 2\nu\sigma \left(\frac{a}{r}\right)^2 \cos 2\theta$$

In particular, for $a = r$ and $\theta = 0$, $\sigma_z = 2\nu\sigma$. This is one of the few problems for which a three-dimensional solution is available to indicate the thickness required to achieve plane strain. It has been shown [8] that a thickness $B \geq 4a$ is required for plane strain to be reached at the centerline of the plate.

If the hole cannot be treated as small compared to the dimensions of the plate, the analysis by way of stress functions is much more difficult. As the hole increases in size, the stress concentration factor (maximum local stress ÷ average net section stress) decreases. However, the value expressed (maximum local stress ÷ average gross section stress) increases so the basis on which stress concentration factors are quoted must be notably difficult. Solutions have been obtained for hole diameters up to one half the width of the plate. When the hole diameter, for example, is 1/3 of the plate width, the maximum stress becomes 3.4S.

Elastic solutions may be superimposed except where nonlinear behavior occurs (contact stresses, buckling). Thus, if 3 and −1 are the maximum and minimum stress concentration factors at the hole for uniaxial tension, we find for biaxial tension or compression stress, concentration factors of 2 and 2, while for shear, the stress concentration factors are 4 and -4 [10].

3.6 Determination of Stresses Around an Elliptical Crack

A problem of greater interest in fracture studies than the circular hole is that of an elliptical hole in an infinite plate loaded by biaxial tensile stresses σ as shown in Fig. 3.9. The solution is a formidable one and was obtained by Inglis [5] in 1913.

By making use of elliptical coordinates ξ and η sketched in Fig. 3.9 where $x = c\cosh\xi \cos\eta$, $y = c\sinh\xi \sin\eta$, Inglis obtained the following expressions for the stresses:

$$\sigma_\xi = \sigma\sinh2\xi\frac{(\cosh2\xi - \cosh2\xi_0)}{(\cosh2\xi - \cos2\eta)^2}$$

$$\sigma_\eta = \sigma\sinh2\xi\frac{(\cosh2\xi + \cosh2\xi_0 - 2\cos2\eta)}{(\cosh2\xi - \cos2\eta)^2}$$

$$\tau_{\xi\eta} = \sigma\sin2\eta\frac{(\cosh2\xi - \cosh2\xi_0)}{(\cosh2\xi - \cos2\eta)^2}$$

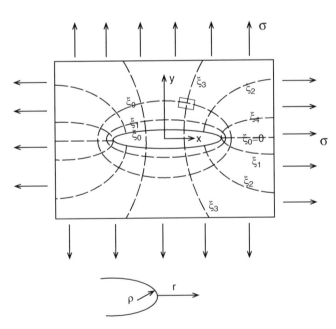

Fig. 3.9 Coordinates used with elliptical crack

The coordinate ξ_0 is the boundary of the crack, while $\eta = 0, \pi$ represent the ends of the major axis. The foci of the ellipse are $x = \pm c$, and the semimajor and semiminor axes are given by $a = c\cosh\xi_0$ and $b = c\sinh\xi_0$. As $\xi_0 \to \infty$, the ellipse becomes a circle, while as $\xi_0 \to 0$, the ellipse becomes a sharp crack of length $2a = 2c$. If we examine the stresses along the x axis (the prolongation of the crack), we find $\sin 2\eta = 0$, $\cos 2\eta = 1$, $r_{\xi\eta} = 0$, and

$$\sigma_x = \sigma_\xi = \sigma\sinh2\xi\frac{(\cosh2\xi - \cosh2\xi_0)}{(\cosh2\xi - 1)^2}$$

$$\sigma_y = \sigma_\eta = \sigma\sinh2\xi\frac{(\cosh2\xi + \cosh2\xi_0 - 2)}{(\cosh2\xi - 1)^2}$$

It may be shown that the stress σ_η for $\eta = 0, \pi$, $\xi = \xi_0$ is given by $\sigma\left(1 + \frac{2a}{b}\right) = \sigma\left(1 + 2\left[\frac{a}{\rho}\right]^{1/2}\right)$ where ρ is the radius of curvature at the tip of the crack. Now letting $\xi_0 \to 0$ yields $x = a\cosh\xi$ to obtain the stress $\sigma_y = \sigma_\eta$ along the x axis,

$$\sigma_y = \sigma_\eta = \sigma\frac{\sinh2\xi}{\cosh2\xi - 1}$$

We can make use of the expressions $x = c\cosh\xi\cos\eta$ and $a = c\cosh\xi_0$ which as $\xi_0 \to 0$ and $\eta = 0$ or π yields $x = a\cosh\xi$ to obtain the stress $\sigma_y = \sigma_\eta$ along the x axis,

$$\sigma_y = \sigma\left(\frac{a}{2r}\right)^{1/2}\frac{1 + (r/a)}{\{1 + (r/2a)\}^{1/2}}$$

where r is the distance along the x axis measured from the crack tip as shown in Fig. 3.9. In the case when $r \to 0$,

$$\sigma_y \simeq \sigma\left(\frac{a}{2r}\right)^{1/2} = \frac{\sigma(\pi a)^{1/2}}{(2\pi r)^{1/2}} = \frac{K_I}{(2\pi r)^{1/2}}$$

Thus, we see that for the infinite plate under biaxial tension, the stress intensity factor is $K_I = \sigma(\pi a)^{1/2}$.

As we saw earlier, this result also applied to a plate loaded by uniaxial tension when the stress is applied perpendicular to the plane of the crack. Similarly, we can show that the stress σ_x along the x axis (i.e., σ_ξ for $\eta = 0$) is

$$\sigma_\xi = \sigma_{x(y=0)} = \sigma\sinh2\xi\frac{(\cosh2\xi + \cosh2\xi_0 - 2)}{(\cosh2\xi - 1)^2}$$

For a sharp crack $\xi_0 = 0$, we have

$$\sigma_{x_{(y=0)}} = \sigma \frac{\sinh 2\xi}{(\cosh 2\xi - 1)}$$

This is the same as the expression obtained for $\sigma_{y_{(y=0)}}$ confirming the results of Eq. (3.1) where for $\theta = 0$ (i.e., $y = 0$), we find $\sigma_x = \sigma_y$. However, the solutions for $\sigma_{x_{(y=0)}}$ for the line crack and the ellipse differ greatly if we consider sharp ellipses other than the extreme case $\xi_0 = 0$. Here we find as in Fig. 3.10 that the stress σ_x at the tip of the crack is zero and rises to a value less than that of the peak σ_y stress at a small distance ahead of the crack tip. At a later stage, we will discuss the delamination of composite materials ahead of a crack, a situation in which the σ_x stress plays an important role. Whether we consider the Westergaard line crack or the Inglis elliptical crack with semiminor axis shrunk to zero, the stress at the tip of the crack is infinite even for vanishingly small loads. Thus, the elastic solutions by themselves are not adequate for discussing fracture. Basically, two approaches have been taken in fracture calculations to overcome this dilemma. One is based on energy considerations since the strain energy associated with the stress field around a crack is finite because the stress is infinite in an infinitesimally small region. The other approach is based on the elastic stress field but makes use of the fact that processes are operating at the tip of the crack which eliminate the infinite stress. It is perhaps confusing to have two different approaches, but they can be shown to be equivalent—at least for fracture in tension. Since much of the literature has been based on these two approaches, we will develop them separately and then show that they are equivalent.

3.7   Elastic–Plastic Solutions

The linear elastic results predict very high stresses at the crack tip even for vanishingly applied loads, and plastic flow will always occur there. The ways in which the stress and strain distributions at a crack tip are affected by this plastic flow can be determined only by performing the appropriate elastic–plastic analysis. This has been done for Mode III cracks [6, 7], but the Mode I crack in plane strain is a much more difficult mathematical problem, and a complete solution is still awaited. If such a crack is assumed to remain sharp after plastic flow occurs [9] except for the small geometry changes associated with a small strain analysis, there is still a singularity at the crack tip; for a nonhardening material, only the strains tend to infinite values. The nature of the singularity has been explored analytically [8, 9], and the results suggest, surprisingly, that there is no significant amplification of strain in the plane of the crack near the tip. The stress σ_y in this region is about $3\sigma_y$ even in a nonhardening material and can reach very high values with strain hardening. The shape of the plastic zone is shown schematically

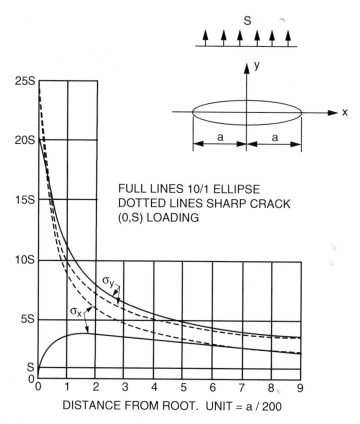

Fig. 3.10 Stress distribution around an elliptical crack

Fig. 3.11 Plastic zone shapes at mode 1 crack tip in plane strain

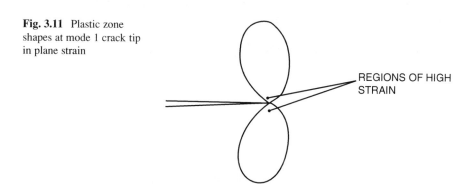

in Fig. 3.11, and the high strains occur in the regions indicated above and below the crack tip. Attempts to obtain elastic–plastic solutions by numerical means, using either finite difference or finite element formulation [11], usually run into difficulties with the high stress and strain gradients at the crack tip, but give useful indications of plastic zone shapes and sizes.

It is clear now from recent results obtained by Rice and Johnson [12] that near-tip solutions based on a sharp crack and small geometry changes give misleading impressions of the near-tip stress and strain fields.

If blunting of the crack tip is taken into account, high strains are found in a small region ahead of the crack, and the stresses do not stay at the high values predicted by sharp tip analyses. The Rice and Johnson solution is an approximate one applicable only when the plastic zone is small compared to the crack length (referred to as small scale yielding). However, finite element configurations have been developed [13] which can in principle cope with the crack tip gradients and geometry changes; it is a matter of time before numerical solutions are obtained to check the predictions about the effects of blunting and to extend the results to large-scale yielding. In further discussion, we will concentrate on the Rice and Johnson analysis as it does seem to contain the important features of the elastic–plastic crack tip problem for plane strain in the opening mode.

The analysis is based on the concept that when the plastic zone is small compared to the crack length and other dimensions such as distance to other boundaries, it is reasonable to suppose that:

(a) If any two cracks in the same material have the same stresses and displacements in the elastic region surrounding the plastic zone, the stresses and deformations within the plastic zone will be the same.
(b) The stresses and displacements in the surrounding elastic region will be given to a close approximation by the linear elastic solution with the value of the stress intensity factor K_I appropriate to the particular crack configuration and applied load.

Elastic–plastic results for Mode III cracks support these assumptions.

The linear elastic solution is thus assumed to provide boundary conditions for the plastically deforming region at the crack tip. It is further assumed that in the first stages of plastic flow, when the tip is sharp, the lines of maximum shear strain in the plastic region are those for a nonhardening material, as plastic flow proceeds the crack tip blunts. It is assumed to blunt to a semicircular arc having diameter δ, as shown in Fig. 3.12. The expression obtained for δ is $\delta \simeq \sigma K_I^2 / ES_y$.

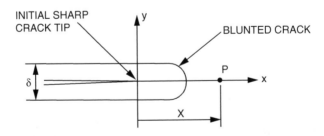

Fig. 3.12 Blunting of crack tip

In the results that are given, attention is focused on the deformations in the plane of the crack. Because the distance from a point in this plane to the crack tip changes with deformation, the results are given in terms of distance X from the position of the initial sharp crack tip to the position of a material point P before deformation—Fig. 3.12. The logarithmic or true plastic strain in the y direction, ε_y^p, is shown as a function of X/δ in Fig. 3.13; there is no plastic strain at distances X greater than about 1.8δ. Figure 3.13 is replotted in Fig. 3.14 to show how ε_y^p at $X = 25, 50, 100$, and 200 ($1\ 1\ \mu m = 10^{-3}$ mm) varies with $\delta(\mu)$. The stress σ_y in the plane of the crack can also be expressed as a function of X/δ and is included in Figs. 3.13 and 3.15 that shows how it varies with δ at different X. At small openings, the near-tip value of σ_y is about $3\sigma_y$, but it is reduced to about $1.16\sigma_y$ by bunting.

Since the plane strain solution assumes $\sigma_z = (1/2)\left(\sigma_x + \sigma_y\right)$, we can determine σ_x in the near-tip region from $\bar{\sigma} = \frac{\sqrt{3}}{2}\left(\sigma_y - \sigma_x\right) = \sigma_y$ or $\sigma_x = \sigma_y - \frac{2}{\sqrt{3}}\sigma_y$. Also, with $\varepsilon_z^p = 0, \varepsilon_x^p = -\varepsilon_z^p$, and thus the effective plastic strain ε^{-p} is given by $\varepsilon^{-p} = \frac{2}{\sqrt{3}}\varepsilon_y^p$.

Fig. 3.13 Stress and strain near blunting mode 1 crack tip in plane strain

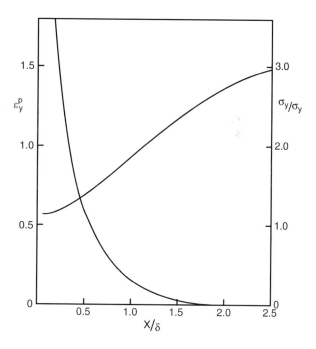

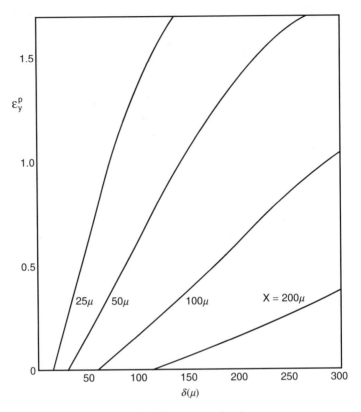

Fig. 3.14 Variation of ε_y^p near crack tip with crack opening δ

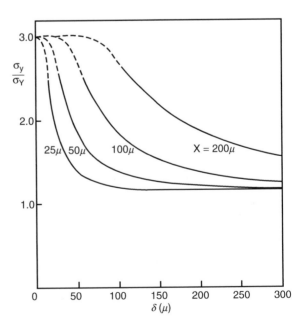

Fig. 3.15 Variation of σ_y/σ_y near crack tip with crack opening δ

3.8   Solutions for Notched Bars

(a) *V-notch bar in bending*

A specimen which has been widely used to study the initiation of cleavage failure in metals is the V-notch bend specimen shown in Fig. 3.16. It is usually thick enough for conditions to be approximately plane strain near the center, and plane strain slip-line solutions given originally by Green [14] are used to estimate the stresses under the notch. These solutions depend on the notch angle 2ψ, the ratio of specimen depth b to depth a under the notch [15], and on the manner of loading. The field for pure bending of a specimen with a sharp notch having

$$3.2° < \psi < 57.3° \quad \text{and} \quad b/a > 1.42$$

is shown schematically in Fig. 3.17. Region I, bounded by the circular arcs AB, is nondeforming, and the right-hand side of the specimen, for example, rotates

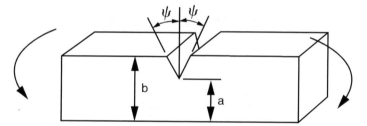

Fig. 3.16 V-notch bend specimen

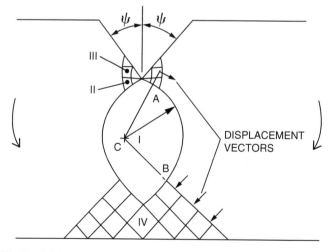

Fig. 3.17 Slip-line field for sharp V-notch specimen in bending $3.2° < \psi < 57.3°$, $b/a > 1.42$

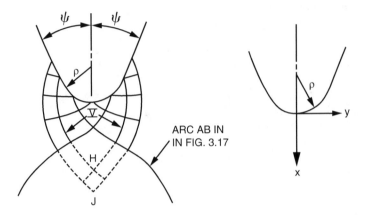

Fig. 3.18 Modification to slip-line field for rounded notch root

about center C with associated outward displacements on the boundaries of regions II and III and inward displacements on IV as indicated.

The modification to the field when the root of the notch is a circular arc of radius ρ is shown in Fig. 3.18. The slip lines in region V are logarithmic spirals, and the stress state in the plane of symmetry is given by

$$\sigma_y = 2k[1 + \log_e(1 + x/\rho)]$$
$$\sigma_x = 2k\log_e(1 + x/\rho)$$
$$\tau_{xy} = 0$$

where x is the distance below the notch—Fig. 3.18. The relative dimensions of the slip-line field depend on the specimen dimensions and are given for one standard specimen in references [16, 17].

The stress field is often extended into the nondeforming region I which can be incipiently plastic [18]. This extension is indicated by the dotted lines in Fig. 3.18, and the stresses are given by the preceding equations down to H which is a distance

$$x_H = \rho^{\left(e^{\frac{\pi}{2} - \psi} - 1\right)}$$

below the notch. Between H and J, the slip lines are straight, and the stresses remain constant at

$$\sigma_y = 2k[1 + \log_e(1 + x_H/\rho)] = 2k[1 + \pi/2 - \psi]$$

$$\sigma_x = 2k[\pi/2 - \psi]$$

For a notch having $2\psi = \pi/4$, for example, the maximum value of σ_y would be 4.36k or $2.18\sigma_y$ for a Tresca yield criterion and $2.56\sigma_y$ for a Mises.

The preceding equations for sharp or slant notches may be used to estimate the stresses in the plastic zone of a partially yielded bar (e.g., provided the position of the elastic–plastic boundary is known or can be determined by etching studies). The estimates would be good in regions where the plastic strains are large compared to elastic strains, but could be in some error near the elastic–plastic boundary. This has been demonstrated recently [19] in an interesting comparison of the stresses predicted by the slip-line solution with those predicted by three different finite element programs for partial yielding of a V-notch specimen in pure bending. Referring to Figs. 3.17 and 3.18, the dimensions of the specimen were

$$a = 8.5\,\text{mm}; \quad b = 12.7\,\text{mm}; \quad \psi = 22.5° \quad \text{and} \quad \rho = 0.25\,\text{mm}$$

and results were given for an applied moments six times that at which plastic flow first occurred. The three programs gave very similar results, and the position of the elastic–plastic boundary and the stresses σ_x and σ_y in the central plane of the notch are shown approximately in Fig. 3.19 for a Mises yield condition and a nonhardening material. The stresses from the preceding equations are also shown and overestimate the maximum value of σ_y by about 14 %. The maximum stresses predicted by the finite element solutions do not occur at the elastic–plastic boundary but (for this particular load) about half way between it and the notch surface.

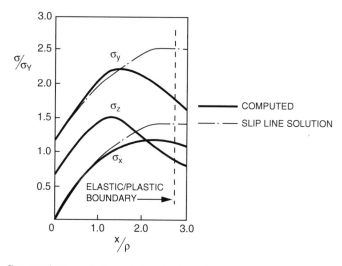

Fig. 3.19 Computed stresses below notch of bend specimen

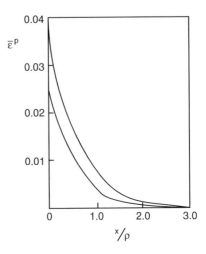

Fig. 3.20 Computed effective plastic strain below notch of bend specimen

The distribution of effective plastic strain in the central plane is shown for the same load in Fig. 3.20. The two lines are the envelopes of results obtained by the three programs.

(b) *Circumferentially notched round bar in tension*

The circumferentially notched round bar in tension is another specimen used in fracture studies. A complete solution for the stress and strain distribution is still being actively sought; in the meantime, the results of an approximate analysis made by Bridgman [20] or the neck of a tensile specimen have been used to estimate the stresses and effective plastic strain when plastic flow is well established over the whole of the minimum cross section of the notched specimen. In this approximate solution, the effective plastic strain ε^{-p}, and hence the effective stress $\bar{\sigma}$, is assumed to be constant over the minimum cross section, and the stresses σ_r, σ_θ, and σ_z are given by

$$\sigma_z = \bar{\sigma}\left[1 + \log_e\left(\frac{a^2 + 2aR - r^2}{2aR}\right)\right]$$

$$\sigma_r = \sigma_\theta = \bar{\sigma}\left[\log_e\left(\frac{a^2 + 2aR - r^2}{2aR}\right)\right]$$

The notch dimensions a and R, and the coordinate system, are defined in Fig. 3.21, and the stresses are plotted in Fig. 3.22. The effective plastic strain is $\varepsilon^{-p} = 2\log_e(a_0/a)$ where a_0 is the initial radius of the minimum section and a the current radius.

The ratio a/R changes during a test. Experiments have shown [21] it to increase with deformation if it is initially less than about 1; if it is initially greater than about 2, it decreases. Between about 1 and 2, there is little change.

Fig. 3.21 Circumferentially notched tension specimen

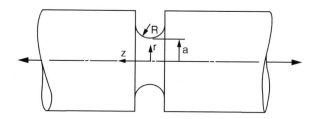

Fig. 3.22 Bridgman solution

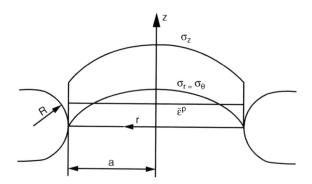

Analyses that have so far been performed to check the Bridgman predictions have given conflicting results (e.g., references [22, 23]). It seems likely, however, that for the sharper notches ($a/R > 1$, say), the effective plastic strain will not be uniform over the cross section but will be smaller at the center of the specimen than at the notch surface. There are finite element programs available which can be used to perform the required analysis, but to obtain solutions for large deformations is very demanding on computer time.

(c) *Plane strain notch tension specimen*

A deeply notched plane tension specimen with a circular notch root is shown in Fig. 3.23; it is assumed to be thick enough for plane strain conditions to apply near the center. One slip-line field which has been used for this specimen [24] is shown in Fig. 3.24 and is the Prandtl field modified to take account of the angled flanks and rounded notch root. Another proposed field [25] is bounded by the line *ABC* shown in Fig. 3.24; both fields give the same stress distribution in the minimum cross section.

In the logarithmic spiral region I, the stresses are given as before by the equation in Sect. 3.8 where x is the distance below the notch root as indicated. The uniform stress state in the central region II is easily obtained in terms of the notch angle 2ψ (Fig. 3.25):

Fig. 3.23 Plane notched
tension specimen

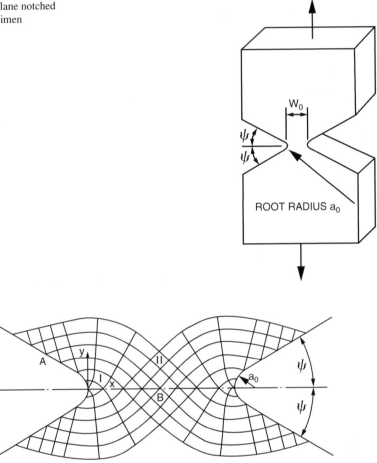

Fig. 3.24 Plane strain slip-line field for notch tension specimen

$$\sigma_y = 2k[1 + \pi/2 - \psi]$$
$$\sigma_x = 2k[\pi/2 - \psi]$$
$$\tau_{xy} = 0$$

which are the same as the equations for the bend specimen. In Fig. 3.25 ΔL is the overall extension of the specimen. Neimark [24] has given results for the strain ε_y^p at the center of the specimen and at the notch root. These depend on the notch angle $2\psi = 90°$ and for $a_0/W_0 = 0.15$ and 0.25, respectively. For $a_0/W_0 = 0.15$, the strain at the notch root is greater than at the center of the specimen; for $a_0/W_0 = 0.25$, it is smaller. The reader is referred to discussions by McClintock [26, 27] on the design of such specimens for fracture studies and on questions of the uniqueness of the strain distributions.

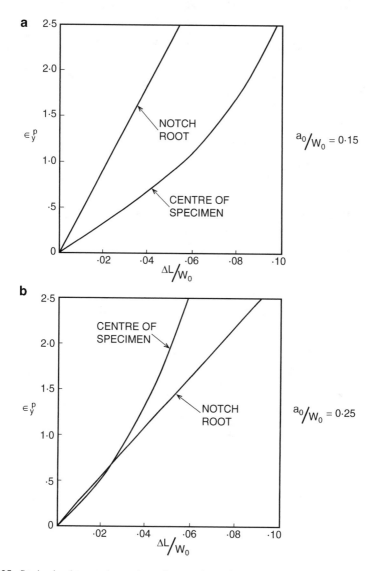

Fig. 3.25 Strains in plane strain notch tension specimen, $2\psi = 90°$

Problems

1. By examining the solution for a sharp elliptical crack of length 2a in an infinite
plate loaded by biaxial tension, you are asked to deduce the distance r/a from the
crack tip at which the expression

$$\sigma_{y(\theta=0)} = \frac{K_1}{(2\pi r)^{1/2}}; \quad K_1 = \sigma(\pi a)^{1/2} \quad \text{is in error by 10\%.}$$

(Note: Quite different results may be obtained for other cracked configurations.)

Stress intensity factors may be obtained experimentally by making use of photoelastic coating. With this technique, one sees a series of fringes (the isochromatics) around the crack tip. Each fringe traces out a line along which the maximum shear stress has a constant value in the x–y plane.

2. (a) You are asked to obtain an expression for the maximum shear stress (in the x–y plane) for a crack under Mode I loading.
   (b) Sketch the general fringe pattern that you would expect to see. Notice that the spacing between fringes corresponds to a constant difference in maximum shear stress.
   (c) Can you recommend a method of plotting the data to obtain K_I?

3. For a Mode I crack, the stress in the z direction is a principal stress. You are asked to determine the principal stresses in the x–y plane near the tip of the crack. What value of θ, for a given r, gives the maximum principal stress and what is its direction relative to the x axis?

   For the same crack, if we examine the stresses near the tip in polar coordinates, for what angle is the stress σ_θ a maximum?

4. For a thin plate under plane stress conditions, it has been suggested that the stress intensity factor for a through crack could be deduced from measurements of the decrease ΔB in the plate thickness B in a region near the crack. Measurements on a material with $E = 69$ GN/m^2, $\nu = 0.3$, $S_y = 350$ MPa show the following data:

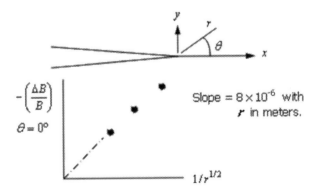

Neglecting the presence of a plastic zone, you are asked to estimate K_I.

5. For a Mode I crack, you are asked to determine the region around the crack tip within which the stresses (calculated from the elastic solution) exceed yield. Use the maximum shear stress criterion for yielding, and sketch your results on dimensionless x, y coordinates

$$\frac{x}{\left(K_1/\sqrt{\pi S_y}\right)^2}, \quad \frac{y}{\left(K_1/\sqrt{\pi S_y}\right)^2}$$

Carry out calculations for both plane stress $\sigma_z = 0$ and plane strain $\varepsilon_z = 0$, $\sigma_z = \nu(\sigma_x + \sigma_y)$, and $\nu = 0.3$.

References

1. Westergaard HM. Bearing Pressure and Cracks. J Appl Mech Trans ASME. 1939(6);A49–53.
2. Muskhelishvili NI. Some basic problems of the mathematical theory of elasticity. Groningen, The Netherlands: P. Noordhoff Ltd; 1953. published originally in Russian in 1933.
3. Williams ML. On the stress distribution at the base of a stationary crack. J Appl Mech. 1957;24:109–14. Trans. ASME, Vol. 79.
4. The solution is originally due to G. Kirsch, VDI, Vol. 42, 1898. It is described in detail in Timoshenko and Goodier's "Theory of Elasticity."
5. Greaves RH, Jones JA. The effect of temperature on the behaviour of iron and steel in the notched-bar impact test. J Iron and Steel Inst. 1925;112:123–65.
6. Hult JA, McClintock FA. Proc. 9th Int. Con. App. Mech., Brussels. 8; p. 51 (1956)
7. Rice JR, Contained plastic deformation near cracks and notches under longitudinal shear. Int. J. Fract. Mech., 1966;2:426–447.
8. Rice JR and Rosengren GF, Plane Strain Deformation Near a Crack in a Power Law Hardening Material. Journal of the Mechanics and Physics of Solids, 16, 1968, pp. 1–12.
9. Hutchison AA, J Mech Phy Solids. Singular behavior at the end of a tensile crack in a hardening material. 1968;(16)13;337
10. Tuba IS. A method of elastic–plastic plane stress and strain analysis. J Strain Anal. 1966;(1) 2:115.
11. Marcal PV, King IP. Elastic–plastic analysis of two-dimensional stress systems by the finite element method. Int J Mech Sci. 1967; 9:143.
12. Rice JR, Johnson MA. The Role of Large Crack Tip Geometry Changes in Plane Strain Fracture, in Inelastic Behavior of Solids (eds. Kanninen MF et al.), McGraw-Hill, N.Y., 1970, pp. 641–672.
13. Levy N, Marcal PV, Ostergren WJ and Rice JR. Small Scale Yielding Near a Crack in Plane Strain: A Finite Element Analysis, International Journal of Fracture Mechanics, 7, 1971, pp. 143–156.
14. Green AP. The plastic yielding of notched bars due to bending. Quart. J. Mech. Appl. Math., Vol. 6, Pt. 2, p. 233 (1953).
15. Ewing DJF. A series method for constructing slip line fields. J. Mech. Phys. Solids, Vol. 16, p. 205 (1968).
16. Alexander JM, Komoly TJ. On the yielding of a rigid plastic bar with Izod notch. Mech. Phys. Solids, Vol. 10, p. 265 (1962).
17. Johnson W. Sowerby R. On the collapse of some simple structures. Int. J. Mech. Sci., Vol. 9, p. 433 (1967).
18. Green AP, Hundy BB. Initial plastic yielding in notch bend tests. J Mech Phys Solids 1956;16:128–44.
19. Owen DRJ, Nayak GC, Kfouri AP, Griffiths JR. Stresses in a partly yielded notched bar—An assessment of three alternative programs. Int. Journ. for Numerical Methods in Eng., Vol. 6, p. 63 (1973).
20. Bridgmen PW. Studies in large plastic flow and fracture. New York: McGraw-Hill; 1952.
21. Alpaugh HE. Investigation of the mechanisms of failure in the ductile fracture of mild steel. S.B. Thesis, Dept. of Mech. Eng., M.I.T. (1965). Results reported by McClintock, F.A. In: Ductility, ASM, Metals Park, OH, p. 255 (1968).
22. Clausing, DP. Effect of plane-strain sensitivity on the Charpy toughness of structural steels. Journal of Materials, Vol. 4, p. 566 (1968).
23. Morrison HL, Richmond O. Large deformation of notched perfectly plastic tensile bars. J Appl Mech. 1972;39(4):971.
24. Neimark, J. E., The fully plastic plane strain tension or a notched bar. J. Appl. Mech., Vol. 35, p. 111 (1968).

25. Hill R, The plastic yielding of notched bars under tension. Quart. Journ. Mech. and Appl. Math., II, Pt. I, p. 40 (1949).
26. McClintock FA. A criterion for ductile fracture by the growth of holes. Int. Jour. Fracture Mechanics, 4, 2, p. 101 (1968).
27. McClintock FA, Liebowitz HA (eds.) Treatise on fracture. Academic Press: New York; 19713:191.

Chapter 4
The Basis of Linear Elastic Fracture Mechanics

4.1 Introduction

In this chapter we will discuss the use of elastic solutions to predict the strength of parts containing sharp cracks. Immediately, we are faced with the difficulty that, whether we consider the line crack or the elliptical crack with semiminor axis shrunk to zero, the stresses at the crack tip are infinite for any value of load. Thus, the elastic solutions by themselves are not adequate for fracture prediction. Basically, two approaches have been taken in fracture calculations to overcome this dilemma without becoming involved in a detailed analysis of the "process zone" at the crack tip. One is based on energy considerations and takes advantage of the fact that the strain energy associated with the stress field around a crack is finite because the stress becomes infinite in an infinitesimally small region. The other makes use of the stress intensity factor to describe the state of stress in the crack tip region. At first sight it is perhaps confusing to have two different approaches but they can be shown to be equivalent.

In discussing the prediction of fracture, we will take a historical route to show how our present understanding of fracture developed. Our starting point is, inevitably, the work of A.A. Griffith (1893–1963), carried out at the Royal Aeronautical Establishment in Great Britain. In 1920 he presented a paper [1] which, quoting Gordon [2], "cut through the mass of tradition and dull detail which hung around materials work but unfortunately no one took him seriously for over 30 years."

Griffith was working on the effect of surface treatments such as filing, grinding, or polishing on the strength of solids. For materials that failed in the elastic range, the approaches to strength prediction used at Griffith's time were that failure occurred at a critical value of either the maximum tensile stress or maximum tensile strain. The stress (or strain) concentration due to a scratch could be estimated from say the solution for an elliptical crack, and on this basis geometrically similar scratches should have the same effect on strength. However,

© Springer Science+Business Media New York 2016
C.K.H. Dharan et al., *Finnie's Notes on Fracture Mechanics*,
DOI 10.1007/978-1-4939-2477-6_4

experimentally, it was known that small scratches were much less damaging than large ones. Thus, either the stress analysis or the failure theory must be wrong. Griffith showed first that large or small scratches do indeed have a similar effect on the local stresses. Long wires with a well-defined yield were given helical scratches with carborundum cloth and then stress was relieved. Under axial load, these spiral scratches made the wire twist. A permanent twist was observed if the axial load exceeded about one quarter of that to produce yielding of the wire as a whole. Since the same result was obtained with coarse and fine scratches, the prediction that geometrically similar scratches should lead to the same local stresses was confirmed. Griffith then concluded that the conventional criteria for predicting fracture must be wrong. This led him to propound an entirely new approach for predicting fracture which formed the basis for much of the subsequent work in this field.

4.2 The Energy Approach

In formulating his criterion for fracture, Griffith took the familiar idea from mechanics that the equilibrium state of a solid body deformed by specified forces or surface displacements is such that the potential energy of the system is a minimum. To this he added the new idea that fracture could only occur if the solid can pass from the unbroken to broken condition by a process involving a continuous decrease in potential energy. The potential energy of the system shown in Fig. 4.1a would normally be taken as the sum of the potential energy of the external load and the elastic strain energy stored in the structure. However, Griffith

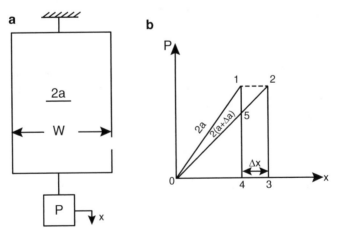

Fig. 4.1 (**a**) Plate of width $w >> 2a$ containing a through crack. (**b**) Load-displacement record as a crack grows under constant load 1–2 or constant displacement 1–5

realized that for fracture calculations another energy term would have to
be considered, that to create fresh surface. The surface energy of a material was a
familiar concept to materials scientists and thermodynamicists at Griffith's time.
It is the energy required to create unit fresh surface due to rearrangement of bonding
forces and, in principle, can be measured in a number of ways unrelated to fracture.
It can be thought of as a thermodynamic property of the material with typical value
of 500–3000 dyn/cm (erg/cm^2) and is denoted by the symbol γ. Griffith realized that
he would have difficulty in treating the formation of small cracks because unless
the surfaces are pulled far apart, compared to the range of the bonding forces, one
would have to know how the bonding forces vary with distance. However, he
pointed out that if a precracked body was considered, in which the crack opening
was large compared to the range of the bonding forces, then the extension of this
crack should result in no great change in the shape of its extremity. The energy
required to create the fresh crack surface can then be taken as the product of 2γ and
the increase in crack area. We will see later that a similar argument about the shape
of the crack tip has to be used in developing the stress intensity approach to fracture
prediction.

Griffith considered the case of an infinite plate of unit thickness containing a
crack of length $2a$. The loading was taken as equal biaxial tension to simplify the
strain energy calculation, but it was deduced that the same result would be obtained
for uniaxial loading perpendicular to the crack plane.

In evaluating the potential energy of the system, only the change in energy
between the uncracked and cracked plate is of interest. Taking the solution for the
elliptical crack with zero semi-minor axis, Griffith found that the strain energy of
the cracked plate exceeded that of the uncracked plate by $v\pi a^2\sigma^2/E$ for the case of
plane stress and by $(1 - v^2)$ times this value for plane strain. In fact this result is in
error as Griffith pointed out in a footnote, and a few years later [3], without
derivation, he gave the correct result $v\pi a^2\sigma^2/E$ for plane stress or $\pi a^2\sigma^2(1 - v^2)/E$
E for plane strain.[1] The source of Griffith's error is interesting since it illustrates a
difficulty that may arise in evaluating the strain energy by integration over the entire
body. As Spencer [4] pointed out, initially Griffith used expressions for the stresses
which were correct at the crack but differed infinitesimally from the true boundary
conditions at infinity. Although providing a satisfactory description of the stress
state, this solution when integrated over an infinite region to obtain the change in
strain energy due to a crack led to significant error.

The decrease in the potential energy of the applied loads when a crack is
introduced or extended under constant load is merely twice the increase in the strain
energy. This can be seen from Fig. 4.1b for the change in the potential energy of the
external loads due to crack growth is $P \cdot \Delta x$ and the increase in strain energy is

---

[1] Griffith's paper [3] shows $(1 - v^2)$ in the denominator.

Fig. 4.2 Components
of energy as a function
of crack length in a plate
with a crack

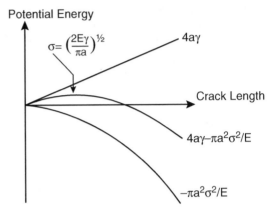

$(1/2)P \cdot \Delta x = \dfrac{\pi a^2 \sigma^2}{E}$. The corresponding increase in surface energy is $4a\gamma$, and
so the total increase in potential energy due to the presence of the crack for plane
stress is

$$U_{\text{cracked}} - U_{\text{uncracked}} = \frac{\pi a^2 \sigma^2}{E} + 4a\gamma - \frac{2\pi a^2 \sigma^2}{E} = -\frac{\pi a^2 \sigma^2}{E} + 4a\gamma$$

This expression is shown graphically in Fig. 4.2, and the maximum in the potential
energy curve is found from $\dfrac{\partial}{\partial a}\left(-\dfrac{\pi a^2 \sigma^2}{E} + 4a\gamma\right) = 0$ or

$$\sigma = \left(\frac{2E\gamma}{\pi a}\right)^{1/2} \tag{4.1}$$

For combinations of stress and crack length which satisfy this equation, the crack
will be in metastable equilibrium. That is, if we imagine an infinitesimal increase in
crack length, the energy released by the external loading and the stress field is just
sufficient to supply the energy required for creation of fresh surface and the
potential energy is unchanged. For cracks larger than specified by Eq. (4.1), a
small increase in crack length will release more energy than can be absorbed in
creation fresh surface. The crack will grow rapidly toward a state of minimum
potential energy with the excess of energy over that required to create fresh surface
appearing as kinetic energy. For a crack length less than that specified by Eq. (4.1),
for a given stress, the minimum potential energy requirement would imply that the
cracks should close up. So in applying Griffith's approach we should specify that
cracks cannot "heal" to prevent them from vanishing. Griffith's result, Eq. (4.1), is a
very important one for it represents the first occasion on which the strength of a
solid was related to the size of the crack which it contains. Similar results are
obtained for plane strain

$$\sigma = \left[\frac{2E\gamma}{\pi a(1 - v^2)}\right]^{1/2} \tag{4.2}$$

and for a penny-shaped crack of radius a with its plane normal to the applied stress [5, 6]

$$\sigma = \left[\frac{\pi E\gamma}{a(1 - v^2)}\right]^{1/2} \tag{4.3}$$

Ignoring the numerical factor which is close to unity, these results are all approximately

$$\sigma = \left(\frac{E\gamma}{a}\right)^{1/2} \tag{4.4}$$

Griffith tested pressurized spheres and cylinders containing cracks and found good agreement between values of surface energy γ deduced from fracture tests and the value of 600 dyn/cm (0.003 lb/in.) obtained by extrapolation from measurements on liquid glass. However, this agreement now appears to have been quite fortuitous.[2] Later, fracture tests on glass [7] in vacuum show values of γ about ten times the thermodynamic value, while for tests in moist air, the corresponding factor is 5. In cleavage tests on certain single crystals at low temperatures [8], the values of surface energy deduced from Eq. (4.1) have been found to agree well with the thermodynamic value, but in general the values of γ deduced from fracture tests are one or more orders of magnitude greater. The reason for this discrepancy is that, normally, dissipative processes are involved in fracture which absorb much more energy than the thermodynamic value considered by Griffith. Even in such an apparently brittle material as aluminum oxide, γ measured from fracture tests at $-196\ °C$ is about 15 times the thermodynamic value (30,000 compared to 2000 ergs/cm^2) which has been attributed to localized plastic deformation [9]. Both in steels where even apparently brittle fracture involves localized plastic deformation [10] and in polymers where long chain molecules are drawn out near the fracture surface [11], the fracture process may require a supply of energy which exceeds the thermodynamic value by a factor of 1000.

---

[2] Several factors may have been involved in this result. First, Griffith's equation was in error by a factor equal to Poisson's ratio. With this correction for $v = 0.26$, his value for γ_{fracture} would be 2400 ergs/cm^2. This compares well with more recent fracture surface energy measurements made over the same loading times (0.5–5 min). In addition, two important factors ignored by Griffith may have been involved. He annealed his specimens after producing the cracks. It has been shown [12] that this may increase strength as much as sixfold relative to unannealed specimens. An opposing factor is that Griffith applied the flat plate solution to cracks in pressurized spheres and cylinders. As we will see later, for the dimensions used by Griffith, this may lead to as much as a threefold decrease in fracture stress relative to flat plate values for a similar crack size.

Despite these quantitative limitations, Griffith's result is still qualitatively useful in that many tests on different materials have shown a stress-crack size relation of the form $\sigma \sim a^{-1/2}$. However, his approach to fracture prediction lay fallow for many years until it was salvaged and extended by Irwin [13] in a series of publications in the late 1940s and 1950s.

Irwin postulated that one could define a surface energy characteristic of fracture which would be measured in a fracture test. He defined a quantity G_c, the work required to create unit increase in crack area by fracture. We note that G_c is twice the value of surface energy deduced from a fracture test for the crack has a top and bottom surface. Since there is no "a priori" reason why G_c should be a material property, this concept at first met a great deal of opposition. In fact, as we will see, G_c may depend on specimen thickness and other test conditions. However, with modifications, Irwin's approach has turned out to be extremely useful and forms the cornerstone of our present-day approaches to fracture prediction. If we assume that G_c, as obtained from a fracture test, is a material property, then, in principle, the load level required to produce fracture in any shape of part with a specified crack may be calculated. This is done by imagining a small increase in crack length and evaluating the energy released by the applied loads and the elastic stress field. Irwin denoted this energy release per unit increase crack area by G_c. Thus, as in the Griffith analysis, if $G < G_c$, the crack is stable; if $G = G_c$, the crack is in metastable equilibrium; and if $G > G_c$, the crack is unstable and should propagate. Although G and G_c have the same units in lb/in^2 or N/M^2, it should be recalled that the first is a quantity calculated from an elastic analysis and the second is a material property. The quantity G_c is often referred to as the "critical energy release rate." G is sometimes referred to as the "crack driving force" (lb/in. of crack front). Alternatively, it may be regarded as the rate of potential energy decrease of the system with crack length.

In practice the calculation of G, by integration over the entire body, can rarely be carried out with a closed form analytical solution. Fortunately, as we will discuss, simpler and equivalent analytical methods are now available and an experimental evaluation of G is also possible. However, with the availability of numerical methods and finite element techniques, the numerical evaluation of G by integrating over the entire body has become a straightforward procedure.

Once solutions for G are available, the critical energy release rate G_c can be determined experimentally. For example, using the infinite plate under uniaxial tension σ as a rather impractical example, since we rarely test such large specimens, we obtain from the Griffith solution $G = \pi a \sigma^2 / E$. The value of G_c is obtained by taking σ and a in this equation as their values at the onset of fracture. We will return later to a more detailed and more practical account of methods for measuring G_c. As pointed out in Chap. 1, its value for different materials, by determining the critical crack size at a given stress level, has a strong influence on the methods of design, inspection, and operation of engineering structures.

An aspect of the energy approach to fracture, which was first raised by Orowan [14] in connection with Griffith's result, is that it is necessary but not sufficient in a mathematical sense. That is, it is not enough to show that an energy condition is satisfied, but also it should be shown that a failure criterion is reached at the tip of the crack. Even though we no longer use Griffith's equation in its original form, it is interesting to examine it from this point of view.

The stress at the tip of an elliptical crack of length $2a$ in an infinite plate subjected to uniaxial tension σ has been shown to be

$$\sigma_{max} = \sigma\left[1 + 2(a/\rho)^{1/2}\right] \cong 2\sigma(a/\rho)^{1/2} \quad a >> p$$

Taking $(E\gamma/b)^{1/2}$ where b is the interatomic spacing as an approximation to the theoretical strength Eq. (4.4), and requiring that this be reached at the tip of the crack, we have

$$\sigma = \left(\frac{E\gamma}{b}\right)^{1/2}\frac{1}{2}\left(\frac{\rho}{a}\right)^{1/2} = \left(\frac{2E\gamma}{\pi a}\right)^{1/2}\left(\frac{\rho}{8b/\pi}\right)^{1/2} \cong \sigma_{\text{Griffith plane stress}} = \left(\frac{\rho}{2.5b}\right)^{1/2}$$

So, if the crack tip radius is about 2.5 times the interatomic spacing, the theoretical strength is reached at the crack tip at the same time as the Griffith energy balance is satisfied. One might conclude that for smaller values of tip radius, the theoretical strength would be reached before the energy condition is satisfied. However, it would be unrealistic to apply an analysis based on a linear elastic continuum to such small regions. About all one can say is that for cracks with tip radii of the order of atomic dimensions, in an ideally elastic material, which is the case Griffith was considering, his result is necessary and sufficient. The fact that the surface energy for fracture is generally much larger than the value taken by Griffith would suggest that the theoretical strength could be reached for somewhat blunter cracks. However, the reason that large surface energies are involved in fracture is that nonelastic processes occur at the tip of the crack. For this reason, the use of an elastic solution to compute the stress at the crack tip is speculative. About all that we can conclude from the preceding discussion is that when cracks become blunt enough, higher stresses will be required to propagate them than specified by an energy balance requirement.

In the modification of Griffith's approach that is now used, the value of the critical energy release rate G_c is found by experiment. Thus we have a necessary and sufficient fracture condition provided G_c is a material property and we work with reproducible cracks.

We have been discussing fracture under constant loads. Such a situation arises with deadweight or pneumatic loading or in a testing machine whose stiffness is small compared to that of the part being tested. It is a simple extension of Griffith's original formulation to treat the case of fracture in a testing machine of arbitrary stiffness. In the extreme case when the testing machine is very stiff compared to the part being tested, the condition is one in which the specimen displacement is controlled during fracture. It is perhaps clearer for this discussion if we regard the specimen as the system and examine the work done on the specimen by the external loading during a small displacement resulting from crack growth. This quantity less the increase in strain energy due to crack growth is the energy available to form fresh surface. Since the work done on the system by the external load results in a corresponding decrease in its potential energy, we are merely transposing a term from one side of the energy balance equation to the other with this procedure.

Referring back to Fig. 4.1b, we see that for fixed load conditions, the work done by the external forces for a crack extension Δa is $P\Delta x$ while the increase in strain energy is $(1/2)P\Delta x$ leaving the balance of $(1/2)P\Delta x$ for the creation of fresh surface. As before, a comparison of $(1/2)P\Delta x$, which by definition is G with G_c tells us whether the crack will or will not propagate. For fixed displacement conditions, the external forces do no work and the decrease in strain energy is the Area 015 in Fig. 4.1b of $(1/2)P\Delta x$. This differs from the energy available for the fixed load case by $(1/2)\Delta x P \Delta x$ which can be neglected since Δx and ΔP are small. Thus, we conclude that, at the moment of crack growth, the rate at which energy is released is the same for fixed load or fixed displacement. Hence, any expression derived for the stress to cause crack growth under fixed load conditions will also apply to fixed grips or any intermediate type of loading condition as sketched in Fig. 4.3. However, as the crack grows, more energy will be released for fixed load than fixed grips. This is evident if we compare hydrostatic and pneumatic burst tests on pressure vessels.

A generalization of the preceding discussion leads us to a very useful result. With reference to Fig. 4.4, we can write the energy released by crack growth Δa as

$$G \cdot B\Delta a = P\Delta x - \Delta U_e$$

where ΔU_e is the change in the elastic strain energy due to crack growth Δa. In the limit we may rewrite this as

$$G \cdot B = P\frac{dx}{da} - \frac{dU_e}{da}$$

Fig. 4.3 Effect of loading conditions on energy release

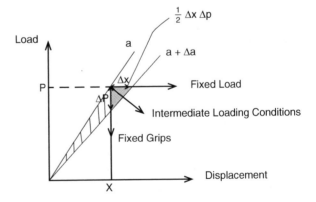

Fig. 4.4 Energy release in terms of dc/da

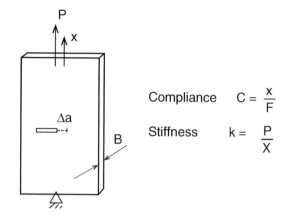

Compliance $C = \dfrac{X}{F}$

Stiffness $k = \dfrac{P}{X}$

Defining the compliance of the cracked body by $c = x/P$, the strain energy is given by $U_e = cP^2/2$ and the preceding equation becomes

$$G \cdot B = P\frac{d(cP)}{da} - \frac{d}{da}\left(\frac{cP^2}{2}\right) = P^2\frac{dc}{da} + cP\frac{dP}{da} - c\frac{PdP}{da} - \frac{P^2}{2}\frac{dc}{da}$$

That is,

$$G = \frac{P^2}{2B}\frac{dc}{da} = -\frac{x^2}{2B}\frac{dk}{da} \quad (G : \text{obtained experimentally}) \qquad (4.5)$$

where $k = P/x = 1/c$ is the stiffness. This result shows that the rate at which energy is released is independent of the loading conditions and depends on only the load, thickness, and change of compliance with crack length. It can be applied to multiple loads in which case

$$G = \sum_i \frac{P_i^2 \, dc_i}{2B \, da}$$

where $c_i = x_i/P_i$, the x_i being the work absorbing displacement under P_i. Thus if P_i is taken as a moment, the corresponding x_i is an angular rotation.

Equation (4.5) opens up the possibility of estimating G by experiment. By progressively extending a crack, e.g., with a saw cut, and making measurements of compliance as a function of crack length, the quantity (dc/da) can be obtained for a part of any shape. This can be done with small loads so that fracture does not occur during the "compliance calibration." Once dc/da as a function of a has been evaluated for a given material, the result may be applied to other materials by adjusting for the different elastic moduli. It is also possible to obtain dc/da from measurements of load and deflection during fracture tests in which several values of fracture initiation and arrest are observed in a single specimen.

Fig. 4.5 Double cantilever
beam (DCB) or trouser legs
specimen

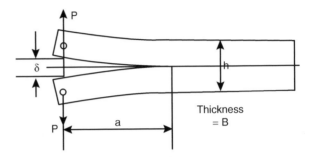

Another useful aspect of Eq. (4.5) is that the compliance may sometimes be estimated from simple "strength of materials" types of solutions. For example, consider the "trouser legs" or double cantilever beam (DCB) specimen shown in Fig. 4.5. Treating each leg as a cantilever of length a, we have from beam theory

$$\frac{\delta}{2} = \frac{Pa^3}{3EI} \quad \text{where } I = \frac{1}{12}B\left(\frac{h}{2}\right)^3$$

Thus

$$c = \frac{\delta}{P} = \frac{2a^3}{3EI}, \quad \frac{dc}{da} = \frac{2a^2}{EI}$$

and the energy release rate is

$$G = \frac{P^2}{2B}\frac{dc}{da} = \frac{P^2 a^2}{BEI} \tag{4.6}$$

In a stiff machine, with δ the controlled variable, the load P decreases as the crack advances. Thus, a number of crack initiations and arrests can usually be observed if side grooves are used to control the direction of the crack propagation. However, we should caution that for many of the DCB specimens used in practice, Eq. (4.6) is not sufficiently accurate. Compliance estimates which consider shear deflections and rotation of the "built-in" end of the leg will be discussed in a later section. Similarly for a DCB specimen loaded by end moments M, we may show that $G = M^2/BEI$. Another case is the double-torsion specimen shown in Fig. 4.6. To find the compliance we note that the deflection y is due to a rotation θ under torque $PW/2$. Since $y = \theta W$ and $\theta = (\text{torque})\frac{a}{GI_p}$ where G is the shear modulus and I_p the polar moment of inertia of a single leg,

$c = \dfrac{y}{P} = \left(\dfrac{W^2 a}{2GI_p}\right)$	$M = EI\dfrac{d^2y}{dx^2} \rightarrow EI\dfrac{dy}{dx} = Mx \rightarrow EIy = \dfrac{Mx^2}{2} \rightarrow \dfrac{\phi}{2} = \dfrac{dy}{dx} = \dfrac{Ma}{EI}$
$G = \dfrac{P^2}{2B}\left(\dfrac{W^2}{2GI_p}\right)$	$\rightarrow c = \dfrac{\phi}{M} = \dfrac{2a}{EI} \rightarrow \dfrac{dc}{da} = \dfrac{2}{EI} \rightarrow G = \dfrac{M^2}{2B}\dfrac{dc}{da} = \dfrac{M^2}{BEI}$

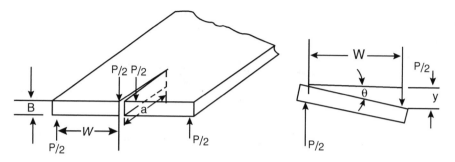

Fig. 4.6 Double-torsion specimen

An interesting feature of the last two cases is that the rate of energy release is not a function of crack length. As we will see, such a result may also be obtained with the DCB specimen shown in Fig. 4.5 if the height h is increased along the specimen. In situations where the specimen is not easily accessible, as in irradiation or corrosion tests, or when the crack length is not easily observed, as in porous materials, it may be useful to test specimens in which G is not a function of crack length.

A fairly recent development in the energy approach to fracture prediction involves examining an arbitrary region around the tip of the crack rather than the entire specimen. This approach although given originally by Eshelby [15] has its origin for engineering applications in a work done independently by Rice [16]. He defined a path-independent integral J for two-dimensional nonlinear elasticity and hence by analogy for deformation theory of plasticity. We will defer a discussion of the many useful aspects of Rice's J integral in elastoplastic or fully plastic fracture problems to a later section. For the present we will indicate how such a path-independent integral, which reduces to G for the linear elastic case, arises from energy balance considerations for a region surrounding the crack tip.

Figure 4.7 shows an arbitrary curve Γ surrounding the crack tip in a plate of unit thickness. We consider the material within Γ as being loaded by stresses σ_x, σ_y, and τ_{xy} acting on the surface formed by the curve and unit thickness. More concisely, we can represent the loading acting on unit area of the surface by a traction vector $\underline{T}$. The corresponding displacement vector at a point on Γ due to the stress field is denoted by $\underline{u}$. As the crack tip moves, $\underline{T}$ and $\underline{u}$ will, in general, change, and since we are dealing with small increments, the work done on the material inside Γ may be written as

$$\int_\Upsilon \left(\underline{T} \cdot \Delta \underline{u} \right) ds$$

where ds is arc length along Γ and the dot product takes care of the fact that the $\underline{T}$ and $\Delta \underline{u}$ vectors may have different directions. On the crack faces $\underline{T}$ is zero so the crack surface makes no contribution to the integral.

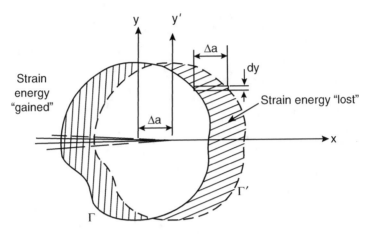

Fig. 4.7 Zone considered in evaluating the J integral

As the crack tip advances by Δa, we can imagine, to a first approximation, the stress and displacement field moving with the crack so that they have the same values with respect to the x'–y' axes as they had before crack growth relative to the x–y axes. The stresses and displacement along Γ' then correspond to the original values along Γ. Thus the displacement increment $\Delta \underline{u}$, on Γ, due to crack growth is given by the change in displacement in moving from Γ' to Γ or to say the same thing in moving Δa from Γ in the $-x$ direction. That is,

$$\Delta \underline{u} = -\frac{\partial \underline{u}}{\partial x} \cdot \Delta a$$

and the work done by the external forces may be written as

$$-\Delta a \int_{\Gamma} \left(\underline{T} \cdot \frac{\partial \underline{u}}{\partial x} \right) ds$$

The change in the strain energy can be evaluated with reference again to Fig. 4.7. As the stress field moves forward, relative to Γ, due to crack growth, the region inside Γ "loses" and "gains" strain energy from the shaded regions. Letting W denote the strain energy per unit volume, the net loss in strain energy is merely the integral of $W \cdot \Delta a \cdot dy$ around Γ. Thus the energy released for unit increase in crack area is

$$J = \int_{\Upsilon} \left\{ W dy - \left(\underline{T} \cdot \frac{\partial \underline{u}}{\partial x} \right) ds \right\} \quad (= G \text{ for the linear elastic case}) \qquad (4.7a)$$

Rice who proved that the J integral is both path independent and also applies to nonlinear elasticity has derived this result more rigorously. Instead of dealing with the vector quantities $\underline{T}$ and $\underline{u}$, we may rewrite the previous result in terms of the stresses σ_x, σ_y, and τ_{xy} and the displacements u and v in the x, y directions

$$J = \int_{\Gamma} \left\{ W\mathrm{d}y - \left[\frac{\partial u}{\partial x}\sigma_x + \frac{\partial v}{\partial x}\tau_{xy} \right]\mathrm{d}y + \left[\frac{\partial v}{\partial y}\sigma_y + \frac{\partial u}{\partial y}\tau_{xy} \right]\mathrm{d}x \right\} \tag{4.7b}$$

The attraction of the energy approach is that one avoids the uncertain details of the stress state at the very tip of the crack. Also, it is possible to include other energy terms in the analysis such as kinetic energy or the energy due to residual stresses. As we will see later, by including a kinetic energy term, estimates have been made for the limiting speed of a crack. Attempts have been made to explain the effect of grain size and irradiation on the strength of beryllium oxide by including the strain energy of residual stresses although later work has cast doubt on this treatment.

The disadvantage of the energy approach is that one has to be very cautious in applying it to more complicated situations such as fracture in compression or under multiaxial stresses or to situations where energy-dissipating processes such as viscoelastic or plastic deformation are present on a large scale. In general, strain energies due to two states of stress can only be directly superimposed if the loads for each state do no work in connection with the displacements of the other. These difficulties were realized by Griffith who was careful to justify that his analysis for equal biaxial tension also applied to uniaxial loading. He pointed out that symmetrical loading relative to the crack had to be involved or else the crack would not propagate in its own plane, thus invalidating the analysis. Perhaps for this reason, in a later paper [3] which treated multiaxial and compressive stresses, he abandoned the energy approach and took as a fracture criterion the maximum tensile stress at an elliptical crack.

4.3 Stress Intensity Approach

As an alternative to the energy approach, we turn now to examine the prediction of fracture using the stress field near the tip of a crack. Since the stress is infinite at the tip of the crack, we cannot equate this to the theoretical strength, but we could, for example, require that the stress in a region of, say, several atomic spacings reach the theoretical strength. Essentially, this is what we did in showing that the Griffith result was sufficient as well as necessary for very sharp cracks. Although Griffith assumed that the crack faces were free of applied loads, he realized that localized stresses of the order of the theoretical strength would exist at the ends of the crack. An approach which incorporates this idea, and which to some is more appealing physically than Griffith's theory, was developed by Barenblatt [17] and others in the (former) Soviet Union. They represent the cohesive forces as an intense force distribution at the ends of the crack as shown in Fig. 4.8a. The external loading

Fig. 4.8 (a) Cohesive forces and smoothly closing curve. (b) Stress intensity factor due to cohesive forces at tip of crack. In this computation it is assumed that the crack faces cannot touch to transmit compression

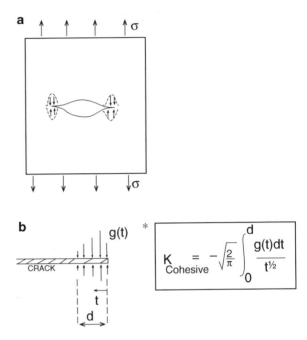

by itself produces infinite tensile stresses at the crack tip while the localized cohesive forces will lead to infinite compressive stresses at the same location.

It is argued that for the cracks to be in what is termed "mobile equilibrium" (metastable equilibrium), they must close smoothly like a zipper. Thus, the combination of stresses due to applied loads and cohesive forces must be such that there is no infinity of stress at the crack tip. Since the only term in the series expansion for the stresses near the crack which goes to infinity is that involving the stress intensity factor, the condition for mobile equilibrium may be written as

$$K_{\text{applied load}} + K_{\text{cohesive forces}} = 0 \qquad (4.8)$$

To develop this concept further, two additional simplifying assumptions are made:

1. The region of the crack acted on by the cohesive forces is very small compared to the crack as a whole.
2. The shape of the crack near its edges, and hence the local distribution of the cohesive forces, does not depend on the shape of the part containing the crack or the applied load. That is, when the maximum intensity, or limit, of the cohesive forces is reached, the crack has a given shape, and when the crack grows, this profile translates in the direction of the crack growth.

The first assumption means that we can calculate the stress intensity factor due to the cohesive forces by treating the crack as a semi-infinite cut in an infinite body, loaded by stresses $q(t)$ for $0 \leq t \leq d$ as shown in Fig. 4.8b. It may be shown that the stress intensity factor is given by

$$K_{\text{cohesive}} = -\sqrt{\frac{2}{\pi}} \int_0^d \frac{q(t)}{t^{1/2}} dt \qquad (4.9)$$

According to the second assumption, this integral will have a constant value, whatever the shape of the part and the magnitude of the loads, at the moment the crack begins to propagate. Thus Eq. (4.8) may be rewritten as

$$K_{\text{applied load}} + K_{\text{cohesive forces}} = \text{constant}$$

This constant which we denote by K_c is considered to be a material property in the same manner as G_c. Of course, we cannot expect to predict K_c from bond strengths just as we cannot use the thermodynamic value of γ for fracture surface energy. However, as stress intensity factors have been computed for many cracked configurations, this approach leads to a very convenient method for fracture prediction. From the figures it can be seen that we have been treating Mode I or direct opening loading but a similar approach can be taken with Modes II and III.

In principle, one only has to measure the value of the stress intensity factor at fracture in one test, and then this may be used to predict fracture in other situations. As an illustrative example, for an infinite plate loaded by biaxial stress σ or uniaxial stresses σ normal to the crack, $K_I = \sigma(\pi a)^{1/2}$. By measuring σ and a at the outset of fracture, we obtain K_c which is usually referred to as the fracture toughness (for Mode I loading). Knowing K_c we can then predict the critical combinations of stress and crack length for any other configuration for which the Mode I stress intensity factor is known. Again as with G and G_c, the distinction between K, the stress intensity factor which provides a description of an elastic stress field, and K_c, the fracture toughness, which is a material property, should be noted.

Before tabulating some stress intensity factor solutions, we should point out a simplification that is often useful. Referring to Fig. 4.9, "A" shows an uncracked body in which stresses $P(x)$ are present on the plane x–x. If now we consider the same part and same loading but with a crack along x–x, as in "B," the original state of stress in A can be obtained by superimposing the stresses in B with the stresses due to crack closing forces $P(x)$ as in "C." The configuration of interest is B and the stress in this case can be obtained by subtracting those in C from A. Thus the stresses in B may be obtained by superimposing the stresses due to condition A and condition C with the sign reversed. However, if we are interested in stress intensity factors, and not in the stresses remote from the crack, case A makes no contribution since it contains no crack. Thus we arrive at the very general result: "The stress intensity factor may be obtained by applying to the faces of the crack the stresses that would exist on the plane of the crack in the uncracked body." This result applies to shear as well as normal stresses allowing K_{II} and K_{III} to be evaluated as

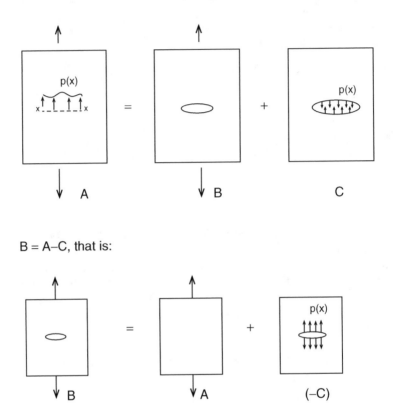

Fig. 4.9 Singularity term for loading B is the same as that for loading $(-C)$

well as K_I. It is useful in simplifying analytical solutions since the stresses for crack face loading go to zero at all boundaries except the crack faces.

As an example, for the Griffith problem, the stress intensity factor is the same as that for a crack loaded by internal pressure σ. We can see from this result that in fracture problems involving pressure vessels, it is important to determine if the fluid can enter the crack. If it can, the stress intensity factor due to the fluid pressure must be added to that due to the applied loads. Such superposition of separate stress intensity factor calculations is one of the great conveniences of this approach to fracture prediction.

Extensive compilations of stress intensity factor solutions are now available [18–20]. In Figs. 4.10, 4.11, 4.12, 4.13, 4.14, and 4.15, we reproduce for future use a few of the many solutions that have been obtained.

Useful general expressions have been given by Barenblatt for the case in which the stress in the plane of the crack $(y = 0)$ in the uncracked body is given in the general form $\sigma_y(x)$. As an example, let us visualize these stresses as arising due to

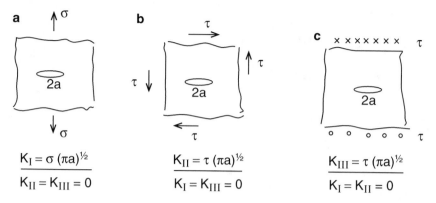

Fig. 4.10 Infinite plate subjected to (**a**) normal stress σ at infinity, (**b**) in-plane shear at infinity, and (**c**) out of plane shear at infinity

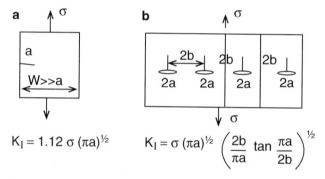

Fig. 4.11 (**a**) Edge crack in infinite plate loaded by normal stress at infinity. (**b**) An array of cracks in an infinite plate loaded by normal stress σ at infinity

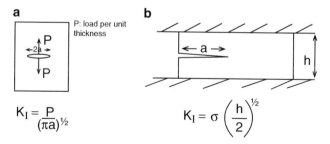

Fig. 4.12 (**a**) Crack in infinite plate loaded by concentrated forces. (**b**) Long strip which is loaded to stress level σ and then held under fixed grip condition while cut a is made

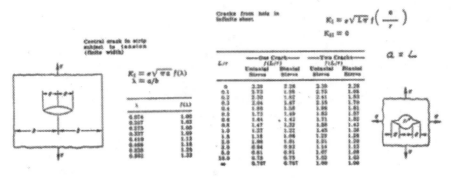

Fig. 4.13 K solutions for which numerical computation is required

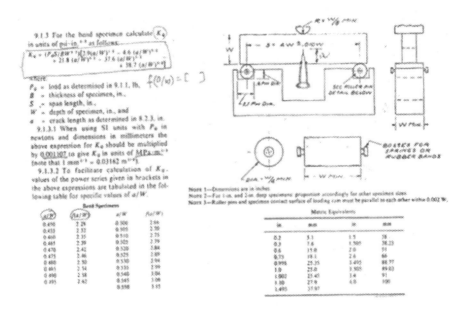

Fig. 4.14 ASTM bend specimen [22]

welding in which case the residual stress distribution $\sigma_y(x)$ will be as shown schematically in Fig. 4.16: then if a crack is introduced which is small compared to the plate, we may write

$$K_1 = \frac{2a^{1/2}}{\pi^{1/2}} \int_0^a \frac{\sigma_y(x)}{(a^2 - x^2)^{1/2}} \, dx \tag{4.9a}$$

for both ends of the crack in Fig. 4.16a.

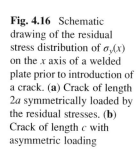
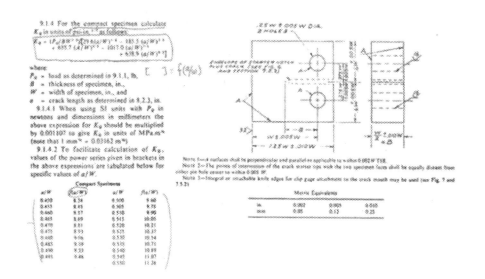

Fig. 4.15 ASTM compact tension specimen [22]

Fig. 4.16 Schematic drawing of the residual stress distribution of $\sigma_y(x)$ on the x axis of a welded plate prior to introduction of a crack. (**a**) Crack of length $2a$ symmetrically loaded by the residual stresses. (**b**) Crack of length c with asymmetric loading

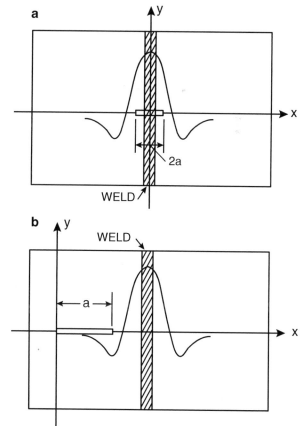

$$K_1 = \frac{2^{1/2}}{(\pi c)^{1/2}} \int_0^a \frac{\sigma_y(x) x^{1/2}}{(c - x^2)^{1/2}} \, dx \tag{4.9b}$$

for the end $x = c$ of the crack in Fig. 4.16b.

Similarly, if we know $\tau_{yx}(x)$ on the plane $y = 0$ prior to inserting the crack, K_{II} may be obtained from equations identical to Eq. (4.9) by replacing $\sigma_y(x)$ with $\tau_{xy}(x)$. We note that the integrals are taken only over the crack length. This is reasonable if we recall the result that stress intensity factors may be found by loading the crack faces with the stress which would exist in the plane of the crack in the uncracked body. As a check, for the case where $\sigma_y(x)$ has the constant value σ, Eq. (4.9a) yields

$$K_1 = \frac{2a^{1/2}}{\pi^{1/2}} \sigma \left[\sin^{-1} \frac{x}{a} \right]_0^a = \sigma(\pi a)^{1/2}$$

Equations (4.9a) and (4.9b) are particularly useful in welding problems since given the residual stress distribution $\sigma_y(x)$ from experiment or analysis, one may compute K_I for different crack lengths and locations. This value of K_I due to residual stresses must be added to that due to the external loads before comparing the total stress intensity factor to K_c for fracture predictions.

The usual application of Eqs. (4.9a) and (4.9b) was inverted by Vaidyanathan and Finnie [21] who used them to determine the residual stresses $\sigma_y(x)$ from measured values of stress intensity factor. As shown by Abel some 150 years ago and in subsequent texts on integral equations, the solution of Eqs. (4.9a) and (4.9b) for σ_y is

$$\sigma_y(a) = \frac{1}{\pi^{1/2}} \frac{d}{da} \int_0^a \frac{K_1(x) x^{1/2}}{(a^2 - x^2)^{1/2}} \, dx \tag{4.10a}$$

$$\sigma_y(c) = \frac{1}{(2\pi c)^{1/2}} \frac{d}{da} \int_0^c \frac{K_1(x) x^{1/2}}{(c - x^2)^{1/2}} \, dx \tag{4.10b}$$

Thus by taking a welded plate, attaching a photoelastic coating, and introducing a cut which is progressively increased by sawing, successive measurements may be made of $K_I(x)$. Using Eqs. (4.10a) and (4.10b), the stresses that existed on the plane of the cut prior to its introduction are obtained. This method of residual stress measurement illustrates that the stress intensity factor has applications in stress analysis which need not involve fracture. The technique of photoelastic coating has also been used to measure stress intensity factors under dynamic conditions with running cracks. Here the measured K_I is assumed to correspond to the dynamic value of K_c.

It is also possible to deduce stress intensity factors from stress concentration solutions [22]. This approach is now somewhat less useful than in the past when there were many stress concentration solutions and few tabulations of stress intensity factors. However, it is of interest in showing the influence of notch tip radius on stress intensity factor.

Typically, we find that stress concentration solutions have the form $\sigma_{max} \simeq 1/\rho^{1/2}$ when the notch depth a or remaining ligament length a with very deep notches is large compared to ρ. For example, for an elliptical crack of length $2a$ in a large plate loaded by uniaxial tension σ,

$$\sigma_{max} = \sigma\left[1 + 2\left(\frac{a}{\rho}\right)^{1/2}\right] \simeq 2\sigma\left(\frac{a}{\rho}\right)^{1/2} \quad \text{if } a \gg \rho$$

as we let $\rho \to 0$ to obtain a sharp crack, $\sigma_{max} \to \infty$, but the product $\sigma_{max}\rho^{1/2}$ is finite. Since the stress ahead of a sharp crack σ_y for $\theta = 0$ is given in terms of the stress intensity factor by

$$\sigma_y = \frac{K_I}{(2\pi r)^{1/2}} \quad \text{or} \quad K_I = \sigma_y \cdot (2\pi r)^{1/2}$$

we might expect to obtain K_I from the stress concentration solution by writing

$$K_I = (\text{constant}) \lim_{\rho \to 0} \left(\sigma_{max}\rho^{1/2}\right)$$

Taking the case we have just discussed for which $\sigma_{max} = 2\sigma(a/\rho)^{1/2}$ and $K_I = \sigma(\pi a)^{1/2}$, we find

$$K_I = \frac{\pi^{1/2}}{2} \lim_{\rho \to 0} \left(\sigma_{max}\rho^{1/2}\right) \tag{4.11a}$$

Similarly it may be shown that

$$K_{II} = \pi^{1/2} \lim_{\rho \to 0} \left(\sigma_{max}\rho^{1/2}\right) \tag{4.11b}$$

$$K_{III} = \pi^{1/2} \lim_{\rho \to 0} \left(\tau_{max}\rho^{1/2}\right) \tag{4.11c}$$

In obtaining these expressions it is understood that only one Mode is present in each case. In Eqs. (4.11a) and (4.11b), σ_{max} is the maximum normal stress adjacent to the notch tip, and in Eq. (4.11c) τ_{max} is the maximum shear stress adjacent to the notch tip.

Since the solution for a sharp deep elliptical notch has the form $\sigma_{max}\rho^{1/2} \simeq 2\sigma a^{1/2}$, we see that the limit in Eq. (4.11a) is reached without having to let ρ approach zero. Thus, we might expect that the stress field around a notch of finite tip radius could be described adequately by the stress intensity factor. We have made this comparison in earlier chapters, and this result justifies the measurement of K values on specimens containing narrow saw cuts. However, the fact that K may be relatively insensitive to crack tip radius does not mean that K_c will be insensitive. By contrast K_c measurements in general show great sensitivity to the shape on the crack tip, and many of the difficulties of fracture toughness testing are associated with the preparation of sharp and reproducible cracks. We have been discussing primarily Mode I stress intensity factors corresponding to through cracks in flat plates subjected to in-plane loading. A more important practical problem is the pressure vessel, and in this case, the bulging that results from the lateral forces of the pressure may lead to much higher stress intensity factors than would be estimated from flat plate solutions. The situation is shown schematically in Fig. 4.17. Not only is the membrane (in-plane) loading increased near the crack tip, which involves a simple multiplicative correction to the flat plate solutions, but also bending loading is introduced. As a result, the stress intensity factor varies across the plate thickness. Solutions have been tabulated for thin-walled cylinders and spheres for these effects in terms of a parameter $\lambda = [12(1 - v^2)]^{1/4} a/(Rh)^{1/2}$ where R is radius, h the wall thickness, and a the half crack length. One has to be cautious with the earlier literature on this topic for some of the solutions were only valid over a limited range of the parameter λ.

Erdogan and Ratwani [23] give a rather complete summary of this topic and we will quote their results. For the direct, or membrane, loading, we have the familiar

Fig. 4.17 Bulging out due
to internal pressure

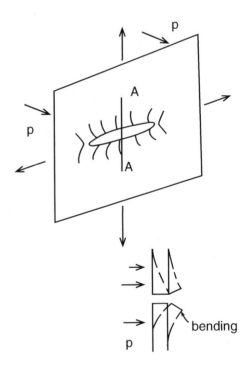

equations from Chap. 3 for the stresses near the crack tip which can be written in the form

$$\sigma_y^m = A_m \frac{K_{1_p}}{(2\pi r)^{1/2}} \frac{1}{4} \left[5\cos\frac{\theta}{2} - \cos\frac{5\theta}{2} \right]$$

$$\sigma_x^m = A_m \frac{K_{1_p}}{(2\pi r)^{1/2}} \frac{1}{4} \left[3\cos\frac{\theta}{2} - \cos\frac{5\theta}{2} \right]$$

$$\tau_{xy}^m = A_m \frac{K_{1_p}}{(2\pi r)^{1/2}} \frac{1}{4} \left[\sin\frac{\theta}{2} - \sin\frac{5\theta}{2} \right]$$

Here K_{1_p} refers to the flat plate stress intensity factor for a through crack, and the coefficient A_m is shown as a function of λ in Fig. 4.18. The stresses near the crack tip due to the bending loading are

$$\sigma_y^b = -A_b \frac{K_{1_p}}{(2\pi r)^{1/2}} \frac{2z}{h} \frac{1}{4(3+v)} \left[(11+5v)\cos\frac{\theta}{2} + (1-v)\cos\frac{5\theta}{2} \right]$$

$$\sigma_x^b = A_b \frac{K_{1_p}}{(2\pi r)^{1/2}} \frac{2z}{h} \frac{(1-v)}{4(3+v)} \left[3\cos\frac{\theta}{2} + \cos\frac{5\theta}{2} \right]$$

$$\tau_{xy}^b = A_b \frac{K_{1_p}}{(2\pi r)^{1/2}} \frac{2z}{h} \frac{1}{4(3+v)} \left[(7+v)\sin\frac{\theta}{2} + (1-v)\cos\frac{5\theta}{2} \right]$$

where A_b is plotted in Fig. 4.18 and the distance z is measured from the neutral axis and is positive when taken inward. If it is assumed that σ_y for $\theta = 0$ dominates in determining fracture when

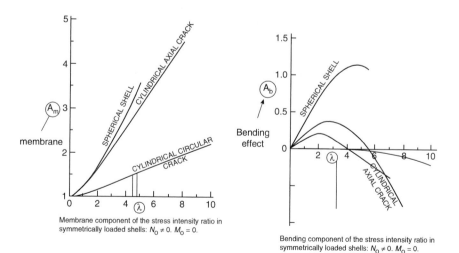

Fig. 4.18 Correction factors for cylindrical shells

$$\sigma_{y,\theta=0} = \sigma_y^m + \sigma_y^b = \frac{K_{1_p}}{(2\pi r)^{1/2}} \frac{2z}{h} \left(A_m - \frac{2z}{h} A_b \right)$$

on the outer and inner surfaces $z = -\frac{h}{2}$ and $z = +\frac{h}{2}$, we have

$$\sigma_{y,\theta=0} = \frac{K_{1_p}}{(2\pi r)^{1/2}} (A_m \pm A_b)$$

with the plus sign referring to the outer and the minus sign to the inner surface. The prediction of fracture from such solutions is far from clear cut. Even after making the assumption that σ_y for $\theta = 0$ controls fracture, we find that this stress varies through the wall thickness. Possibly, in a very brittle solid, the value $(A_m + A_b)K_{1_p}$ could be used to compare flat plate fracture studies with pressurized spheres or cylinders containing through cracks. However, as we see in discussing tests on metals, fracture correlations based on the corrected stress intensity factor $(A_m + A_b)K_{1_p}$ are not entirely satisfactory. The values of A_m and A_b shown in Fig. 4.18 were obtained by Erdogan and Ratwani for a value of Poisson's ratio $= 1/3$. However, these authors also showed that if the correction factors A_m and A_b were plotted as a function of $a/\sqrt{Rh}$, the effect of Poisson's ratio was negligible. As an approximation adequate for design purposes, in the range $0 \le \lambda \le 5$, we can add A_m and A_b in Fig. 4.18 to obtain

	Sphere	Cylinder (axial crack)
$\sigma_y(\theta = 0)/$(Flat Plate Value)	$(1 + 0.75\lambda)$	$(1 + 0.45\lambda)$

Apart from the penny-shaped crack in the infinite solid, Eq. (4.3), all of the cracks we have been discussing are through cracks in flat of curved sheets. The penny-shaped crack is a special case of the more general result [24] for a flat elliptical crack in an infinite solid with its plane normal to the applied stress σ. This solution has proved extremely useful in treating internal flaws, e.g., due to inadequate welding, and has been modified to treat surface or thumbnail flaws. Referring to Fig. 4.19, the stress intensity factor is given as a function of angular location ϕ by

$$K(\phi) = \sigma \left(\frac{\pi a}{q^2} \right)^{1/2} \left\{ \left(\frac{a}{b} \right)^2 \cos^2\phi + \sin^2\phi \right\}^{1/4} \tag{4.12}$$

where q is the complete elliptical integral of the second kind given by

$$q = \int_0^{\pi/2} \left(1 - k^2 \sin^2\alpha \right)^{1/2} d\alpha \quad k^2 = 1 - \frac{a^2}{b^2}$$

We should note a change in notation at this stage since a corresponds to the semiminor axis of the ellipse and b to the semimajor axis. The maximum value of

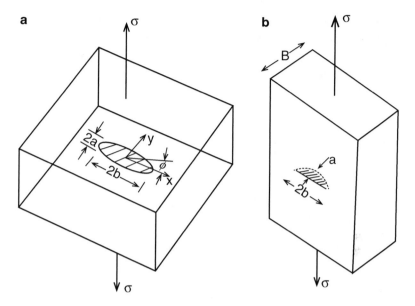

Fig. 4.19 (a) Elliptical flaw in infinite body. (b) Semielliptical flaw in semi-infinite body. The load axis is perpendicular to the plane of the crack

K occurs at $\phi = 90°$ and is $K = \sigma(\pi a/q^2)^{1/2}$. For the case of the penny-shaped crack $(a = b)q = \pi/2$ and $K = 2\sigma(a/\pi)^{1/2}$.

The type of crack very often observed in practice is the thumbnail or semielliptical surface flaw shown in Fig. 4.19. Correction factors have been computed for this case to take care of the effect of the free surface. These range from a multiplying factor on $K(\phi)$ of Eq. (4.12) of 1.122 for $a/b \to 0$ to 1.03 for $a = b$. A suggested value [25] for other values of a/b is $1 + 0.122(1 - a/2b)^2$. Apart from corrections for the effects of local plastic deformation, which we will discuss in the next chapter, the preceding thumbnail crack solution may also require correction if the crack front approaches the back face of the plate (say if $a > B/2$). In addition, allowance may have to be made for the effect of curvature if the factor $\lambda = [12(1 - v^2)]^{1/4}b/(Rh)^{1/2}$ is large enough.

To sum up, the stress intensity approach to fracture prediction appears easier than the energy approach because we only look at a single quantity, the stress intensity factor, instead of evaluating the strain energy for the entire body; however, as we will see, the Griffith energy balance can be recast so that one only has to consider the stress state near the tip of the crack. Although Barenblatt's concepts might appear to describe the fracture process more accurately than Griffith's, this is not really the case for as shown by Cribb and Tompkin [26] the cohesive forces invoked by Barenblatt act on a region of only several atomic spacings.

4.4 The Equivalence of Energy and Stress Intensity Approaches

The use of stress intensity factors to predict the fracture of high-strength metals dates back only about two decades. At that time it was known [10] that the plastic work accompanying fracture in metals exceeded the thermodynamic surface energy by orders of magnitude. However, Irwin's contention that a modified surface energy, G_c, could be used for fracture prediction found little support. This is perhaps not surprising for then, as now, there was no a priori evidence that G_c should be a material property. Also, the analytical difficulties in calculating energy release rates for practical shapes undoubtedly limited the development of the energy approach to fracture prediction.

The stress intensity factor approach to fracture prediction was developed extensively in the Russian literature, and solutions were obtained for many cracked configurations. However, rather surprisingly, the application of this work was largely to problems of a geological nature. Until Irwin [27] showed that the energy and stress intensity factor approaches were equivalent, for linear elastic behavior, there appears to have been no application of stress intensity factors to predict fracture in metals. By showing that the energy release rate in a cracked body depends only on the conditions at the tip of the crack, Irwin unified the two approaches to fracture prediction and stimulated a great deal of activity in the field of fracture.

We will reproduce Irwin's development for the case of plane strain. Fixed grips are assumed for simplicity, although we know that loading conditions will not affect the energy release rate for given stresses.

The energy released by the stress field when a crack grows by a small amount α in the plate of unit thickness shown in Fig. 4.20 is by definition $G\alpha$. Next we examine the stresses, σ_y, and displacements, v, in the y direction on the short length before and after crack extension occurs. From the figures we see that the stress

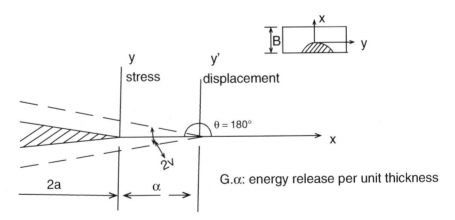

Fig. 4.20 Crack of length $2a$ in plate of unit thickness

changes from the value σ_y with respect to the y axis (for $\theta = 0$) to zero while the displacement grows from zero to a value v with respect to the y^1 axis for $\theta = \pi$.
  Hence

$$G \cdot \alpha = 2 \int_0^\alpha \frac{1}{2} \sigma_y v \, d\alpha$$

where the factor 2 in front of the integral takes care of the upper and lower faces of the crack. From Eq. (3.1) we have for $\theta = 0$, $r = x$, $\sigma_y = K_1/(2\pi x)^{1/2}$, while for $\theta = \pi$, $r = \alpha - x$, $v = \frac{K_1}{G}(\alpha - x/2\pi)^{1/2}(2 - 2v)$.
  Hence

$$G = \frac{K_I^2}{G} \frac{(1-v)}{\pi} \frac{1}{\alpha} \left(\frac{\alpha - x}{x}\right)^{1/2} dx$$

Since

$$\frac{1}{\alpha} \int_0^\alpha \left(\frac{\alpha - x}{x}\right)^{1/2} dx = \int_0^1 \left(\frac{1-\xi}{\xi}\right)^{1/2} d\xi = \frac{\pi}{2}$$

we find that

$$G = \frac{K_I^2(1-v)}{2G} = \frac{K_I^2(1-v^2)}{E} \quad \text{for plane strain} \tag{4.13}$$

Similarly it may be shown that for plane stress $G = K_I^2/E$. So, determination of the critical value of the energy release rate G_c is equivalent to measuring the critical value of the stress intensity factor K_c. Usually, the notation K_{Ic} is reserved for plane strain because plane stress fracture, as we will see, is usually not clear-cut Mode I cracks.
  In the more general case, it may be shown[3] that

$$G = \frac{K_I^2}{E} + \frac{K_{II}^2}{E} + \frac{K_{III}^2}{2G} \qquad \text{plane stress}$$

$$G = \frac{K_I^2}{E}(1 - v^2) + \frac{K_{II}^2}{E}(1 - v^2) + \frac{K_{III}^2}{2G} \quad \text{plane strain}$$

One should not infer from this result that K_{Ic}, K_{IIc}, and K_{IIIc} values are related since the localized energy-absorbing processes at the crack tip may be quite different for the different modes of crack growth. The majority of studies have been made on Mode I cracks, but later we will discuss cases of mixed mode loading. A more detailed comparison of the energy and stress intensity approaches has been given by

---

[3] In other than two-dimensional problems, e.g., the thumbnail crack, we may have to be cautious in comparing a global approach G with a local approach K, when K varies with location.

Willis [28]. He showed that they are identical except for the fact that one involves work to separate two planes from an unstrained position and the other the work to separate from a strained position.

The result which we have quoted, namely, that the energy released by crack extension is merely $1/2 \int$ (stresses prior to crack extension) $\times$ (displacements after extension), with the integral taken over the crack extension only, is only one of a number of useful results which have been obtained in recent years. It has been shown by Bueckner [29] that the energy released by crack growth is equal to the strain energy of an elastic system with stresses and strains equal to the difference between final and initial states. In another very useful formulation, Rice's J integral (Eqs. 4.7a), we have seen that the rate of energy release by crack growth may be expressed in terms of an integral taken over a region surrounding the tip of the crack. Since the region of integration can be made vanishingly small, any stress terms in the integral that do not become infinite at the tip of the crack do not contribute to the rate of energy release due to crack growth. All this opens up many possibilities for calculating G from the stresses at the crack tip alone and shows that the Griffith energy balance depends only on the singularity (infinity) of stress at the crack tip.

4.5   Displacements Due to Cracks

Traditionally, in engineering design, elastic solutions are used to calculate the deflections of uncracked members such as beams and shafts. It is useful also to be able to calculate the additional deflection due to the presence of a crack. This approach which is based on the Theorem of Castigliano as described in Chap. 2 and the relation between K and G has been developed fairly recently.

The strain energy of a cracked body may be written as the sum of the strain energy of the uncracked body plus the change $\overline{U}_e$ in the strain energy due to the presence of the crack.

From Castigliano's theorem the displacement x_i which occurs under a load P_i when a linear elastic body is subjected to loads P_l, $-P_i - P_n$ that give an equilibrium state is

$$x_i = x_i(\text{uncracked body}) + \frac{\partial \overline{U}_e}{\partial P_i}$$

In the subsequent derivations, the displacements in the uncracked body will be neglected. They may be obtained by conventional procedures and are zero if displacements at a crack surface are being considered. For plane problems involving through cracks, we recall that the growth of a crack from a to $a + \Delta a$ in a plate of thickness B results in an energy release.

$G \cdot B \Delta a =$ (Work done by external forces $-$ increase in Strain Energy)$_{\text{due to } \Delta a}$

or $G \cdot B \Delta a = \left(2\Delta \overline{U_e} - \Delta \overline{U_e} \right)$ for fixed loads.

In the limit

$$G \cdot B = \frac{\partial \overline{U_e}}{\partial a} \quad \text{or} \quad \overline{U_e} = \int_0^a G \cdot B \Delta a \quad \text{since} \quad \overline{U_e} = 0 \quad \text{when} \quad a = 0$$

For an axisymmetric crack, the thickness B has to be replaced by the length of the crack front $2\pi a$.

Considering the plane case, the Theorem of Castigliano leads to

$$\left(x_i \right)_{\text{due to crack}} = \frac{\partial}{\partial P_i} \int_0^a G \cdot B \cdot da$$

where G may be written in terms of the stress intensity factors.

As with the conventional application of Castigliano's Theorem, if the displacement is required at a location where no forces are present, fictitious forces are introduced at this location and set to zero in the final result.

To illustrate this, the opening at the center of a crack of length $2a$ in an infinite plate of unit thickness subjected to uniaxial tension σ will be calculated. The complete solution to this problem was given in Chap. 3. Figure 4.23a shows the introduction of fictitious forces P per unit thickness at the location where the displacement is to be determined. From the known stress intensity factor solutions,

$$K_1 = \sigma(\pi a)^{1/2} + \frac{P}{(\pi a)^{1/2}}$$

Since the crack is of length,

$$\overline{U_e} = 2 \int_0^a G \cdot da = \frac{2}{E'} \int_0^a \left[\sigma(\pi a)^{1/2} + \frac{P}{(\pi a)^{1/2}} \right] da$$

$$x_P = \frac{\partial \overline{U_e}}{\partial P} = \frac{4}{E'} \int_0^a \left[\sigma(\pi a)^{1/2} + \frac{P}{(\pi a)^{1/2}} \right] \frac{P}{(\pi a)^{1/2}} da$$

Note that it is simpler to differentiate before integrating and at this stage the fictitious force P may be set to zero. Thus

$$x_P = \frac{4}{E'} \int_0^a \sigma da = \frac{4a\sigma}{E'}$$

Similarly for the edge crack of length a in a semi-infinite plate loaded by uniaxial tension σ, we introduce the fictitious forces F. From computations of stress intensity

factors, $K = 1.12\sigma(\pi a)^{1/2} + \left[2.60F/(\pi a)^{1/2}\right]$. Proceeding as in the previous problem, $\Delta = 5.83\sigma a/E'$. This solution could also be used to estimate the opening of a
surface crack in bending with σ taken as the stress at the outer fiber if the crack
length is very small compared to the width of the beam.

A useful solution for measurements of J_c is the deeply notched plate in bending
shown in Fig. 4.23c. Taking M as the moment per unit thickness, the compilations
of stress intensity factors provide $K_1 = 3.975M/b^{3/2}$. The rotation due to the crack
θ_c is given by Castigliano's Theorem

$$\theta_c = \frac{\partial}{\partial M}\int_0^a G \cdot da = \frac{\partial}{\partial M}\int_\infty^b G(-db) \;\rightarrow\; \theta_c \simeq \frac{\partial}{\partial M}\int_b^\infty \frac{(3.975)^2 M^2}{E' b^3}\,db \simeq \frac{15.8 M^2}{E' b^3}$$

This result is not exact since the stress intensity solution is for a deeply notched
crack and the integration is taken from $b = b$ to $b = \infty$. However, the presence of b^3
in the denominator of the integral shows that contributions to θ_c from large values
of b will be very small and hence the use of an inexact stress intensity factor for this
region should not lead to serious error.

A similar problem arises in treating the deeply notched circular rod under
uniaxial load shown in Fig. 4.23d. For this case $K_1 = P/(2\sqrt{\pi}b^{3/2})$. Remembering
that the axisymmetric problem has a crack front of $2\pi b$,

$$\delta_c = \frac{\partial}{\partial P}\int_b^\infty Gd(\text{Area}) \simeq \frac{\partial}{\partial P}\int_b^\infty \frac{P^2}{4\pi b^3 E'}2\pi b \cdot db \;\rightarrow\; \delta_c \simeq \frac{P}{E' b}$$

As a final example we determine the compliance of the cracked plate shown in
Fig. 4.23e. Because the crack length is $2a$,

$$\overline{U}_e = 2B\int_0^a G da = \frac{2B}{E'}\int_0^a \sigma^2 \pi a da$$

Putting $\sigma = P/BW$,

$$\overline{U}_e = \frac{2\pi}{E' BW^2}\int_0^a P^2 a da$$

$$x_P = \frac{\partial P \pi a^2}{E' BW^2} = \frac{2\pi}{E'}\frac{\sigma a^2}{W}$$

The total deflection is that due to the crack plus $\sigma h/E$ that of the uncracked plate.
Considering plane stress, $E' = E$, the total deflection is $(\sigma h/E)(1 + 2\pi a^2/Wh)$. The
ratio of the compliance c of a cracked plate to that c_0 of an uncracked plate is then

$(c/c_0) = 1 + (2\pi a^2/Wh)$. If n non-interacting through cracks were present, the strain energies due to the cracks can be added and

$$\frac{c}{c_0} = 1 + \frac{n2\pi a^2}{Wh}$$

4.6   Work of Fracture

In this approach, instead of measuring the value of G_c for crack initiation, or its equivalent K_c, the work involved in fracturing a specimen is measured and divided by the area of the fracture surface to estimate G_c. Such an approach may be reasonable if a test configuration is used in which fracture occurs slowly and in equilibrium. In this case a record of load and displacement as shown in Fig. 4.21 may be used to estimate G_c if crack length readings are also taken. In moving from point 1 to point 2 on this curve, the work done by the external force less the increase in strain energy is merely the area of the shaded triangle. Hence an estimate for G_c is obtained by dividing the area of the shaded triangle by the increase in cracked area between points 1 and 2. Several aspects must be considered in comparing work-of-fracture values for G_c to values obtained from crack initiation tests. Conditions for the initiation and the propagation of a crack may vary. For example, G_c may be a function of crack velocity and the mode of fracture may change, e.g., from flat to slant, as the crack propagates. Even for ceramic materials which all show flat fracture on a macroscopic scale, work of fracture has been reported [30] to give higher values of G_c than initiation values for graphite but lower values for glass and PMMA. In metals if a change occurs from flat to a slant surface as the crack propagates away from the starter slot, initiation and propagation values may differ greatly. To overcome some of the limitations of work-of-fracture measurements, the specimen shown in Fig. 4.22 has been used. The work-of-fracture values are measured for different values of d/b, and extrapolation to $d/b = 0$ yields values which appear to be in agreement with DCB tests [31].

Fig. 4.21 Load-displacement record during stable cracking. Crack area increases from A1 to A2 from point 1 to point 2. Energy released by loads and stress field is 1234 +014 − 023 = 012

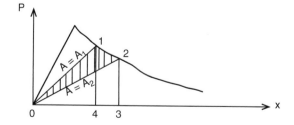

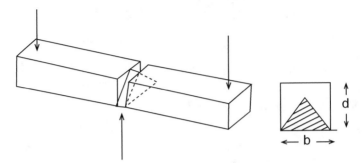

Fig. 4.22 Specimen configuration used in work-of-fracture studies [31]

The traditional "work-of-fracture test," although it is not usually reported in this sense, is the Charpy V-notch impact test. In any dynamic test of this type, it is difficult to separate the work of fracture from kinetic energy unless precise instrumentation is accompanied by careful analysis (Fig. 4.23).

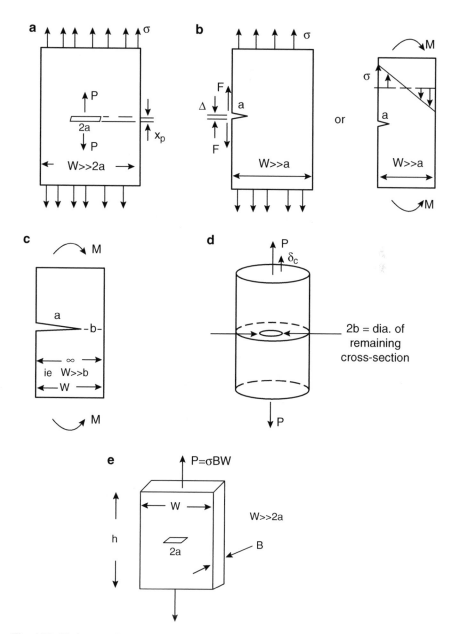

Fig. 4.23 Various crack configurations and loading conditions (see text)

Problems

1. Mica, a transparent almost ideally brittle material, can be cleaved, as shown below, by a smooth rigid wedge. One cannot measure the force P in such an experiment, but the deflection y of the cleaved strip near the crack front can be measured optically by counting interference fringes. Obreimoff [32] made such an experiment and for $h = 0.011$ cm found

$$y = 0.04x^2 \quad \text{where } x \text{ and } y \text{ are in centimeters.}$$

Treating the cleaved strip as a cantilever of length a, and ignoring shear deflections, you are asked to estimate G_c (dyn/cm) for Obreimoff's muscovite mica. We recall that the deflection curve of a cantilever of length a with end load P is given by

$$y = \frac{P}{EI}\left(\frac{ax^2}{2} - \frac{x^3}{6}\right) = \frac{Pax^2}{2EI} \quad \text{if} \quad x \ll a$$

Take E for mica as 2×10^{12} dyn/cm^2.

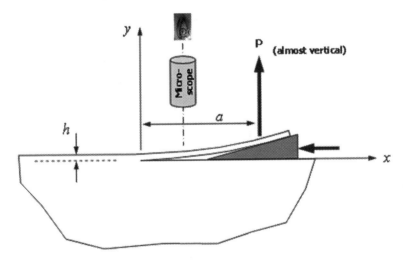

(assume unit width perpendicular to paper).

2. It has been proposed that fracture tests be conducted on double cantilever beam specimens loaded by end moments as shown. Recalling that the end rotation ϕ of a cantilever beam of length a is given by $\phi = Ma/EI$, you are asked to obtain an approximate value for K_I for this specimen for plane strain conditions. Express the result in terms of M and the specimen dimensions.

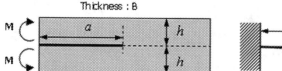
Thickness : B

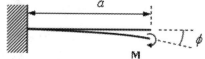

3. To measure the fracture toughness G_c of an adhesive joint, the test sketched below has been proposed. The strain energy of a plate clamped at its rim and centrally loaded is given by

$$U = \frac{8\pi Dx^2}{a^2} \quad \text{where,} \quad D = Eh^3/12(1 - v^2) \quad \text{and} \quad x = \frac{Pa^2}{16\pi D}$$

You are asked to obtain an expression for G in terms of P, D, and any other quantities involved.

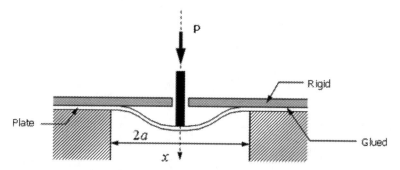

4. A plate, as shown below, is clamped at the ends but no external loads are applied. If the ambient temperature decreases but the grips do not move, then an axial tensile stress will be developed in the plate (transverse stresses can be neglected well away from the grips). For a high-strength steel plate 1 in. thick with the properties given below, you are asked to estimate the temperature change that will cause catastrophic propagation of a 75 mm long through crack:

Young's modulus, $E = 200$ GPa
Coefficient of thermal expansion, $\alpha = 10.8 \times 10^{-6}/°C$
Critical crack energy release rate, $G_{Ic} = 5250$ J/m^2
Yield strength, $S_y = 1400$ MPa
Poisson's ratio, $v = 0.3$

Assume plane strain and no change of G_{Ic} with temperature which may be a reasonable assumption if we are considering moderate changes in temperature. Also neglect any corrections.

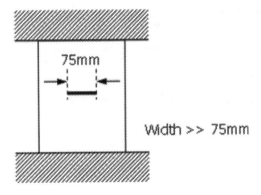

References

1. Griffith AA. Philos Trans R Soc Lond. 1920;A221:163–98.
2. Gordon JE. The new science of strong materials, vol. A920. Harmondsworth: Pelican Books Ltd; 1968.
3. Griffith AA. Proc. 1st Int. Cong. Appl. Mech., Delft, Holland; 1924. p. 53–64.
4. Spencer AJM. Int J Eng Sci. 1965;3:441–9.
5. Sneddon IN. Proc R Soc Lond. 1946;187A:229–60.
6. Sack RA. Proc Phys Soc. 1946;58:729–36.
7. Linger KA, Holloway DG. Philos Mag. 1968;18:1269.
8. Gilman JJ. J Appl Phys. 1960;31:2208–18.
9. Congleton JA, Petch NJ. Acta Metall. 1966;14:1179–82.
10. Orowan E. Fracture and strength of solids. Rep Prog Phys. 1948;12:185–232.
11. Berry JP. J Polym Sci. 1961;50:107–15. 313.
12. Berg CA. Proc. Joint International Conference on Creep, Inst. Mech. Engineering, London; 1963. p. 1.41–1.47.
13. Irwin GR. Fracture dynamics, fracturing of metals. Cleveland: ASM; 1948. p. 147 (see also many later references).
14. Orowan E. Trans Inst Eng Shipbuilders (Scotland). 1945;89:165–215.
15. Eshelby JD. Solid state physics, vol. 3. New York: Academic; 1956.
16. Rice JR. Trans ASME J Appl Mech. 1968;35:379–86.
17. Barenblatt GI. In: Dryden HL, von Karman T, editors. Advances in applied mechanics, vol. 7. New York: Academic; 1962. p. 55–129.
18. Rooke DP, Cartwright DJ. Compendium of stress intensity factors. UK: H.M.S.O.; 1976.
19. Sih GC. Handbook of stress-intensity factors. Bethlehem, PA: Lehigh University; 1973.
20. Tada H, Paris P, Irwin GP. The stress analysis of cracks handbook. Hellertown, PA: Del Research Corp; 1973.
21. Vaidyanathan S, Finnie I. Trans ASME J Basic Eng. 1971;93D:242–6.
22. Paris PC, Sih GC. Fracture toughness testing and applications. ASTM STP. 1965;381:30–83.
23. Erdogan F, Ratwani M. Nucl Eng Des. 1972;20:265–86.
24. Green AE, Sneddon IN. Proc Camb Philos Soc. 1950;46:159–63.
25. Kobayashi AS, Moss WL. Proc. 2nd Int. Conf. on Fracture, Brighton, England. London: Chapman and Hall; 1969.
26. Cribb JL, Tompkin B. J Mech Phys Solids. 1967;15:135–40.
27. Irwin GR. Trans ASME J Appl Mech. 1957;24:361–4.

28. Willis JR. J Mech Phys Solids. 1967;15:151–62.
29. Bueckner HF. Trans ASME. 1958;80:1225–9.
30. Davidge RW, Tappin G. The effective surface energy of brittle materials. U.K. Atomic Energy Authority Report AERE—R5621, 1967.
31. Simpson LA, Wasylyshyn A. The measurement of work of fracture of high strength brittle materials. Atomic Energy of Canada Report AECL—3677, Feb 1972.
32. Obreimoff JW. Proc Roy Soc. 1930;A127:290–7.

Chapter 5
Some Applications of Linear Elastic Fracture Mechanics

5.1 Introduction

The approach to fracture prediction that we discussed in the preceding chapter has been applied to a wide variety of materials in many types of applications. While steels, aluminum alloys, and titanium alloys have received the greatest attention with an extensive data compilation [1] now being available, many other metals have also been studied. In addition, extensive fracture toughness testing has been carried out on such diverse materials as ceramics, rocks, polymers, adhesive joints, and a great variety of composites. Many types of tests have been used to measure fracture toughness in these materials. We will defer out discussion of this topic until we have covered the modifications that have to be made to elastic solutions to cope with localized plastic deformation in the crack tip region.

As with all generalizations, it is difficult to define the range of applicability of linear elastic fracture mechanics to a wide variety of materials. Clearly, we would expect little difficulty in applying elastic solutions to ceramics and high-strength steels. For lower-strength alloys where relatively larger amounts of plastic deformation precede fracture, the linear elastic solutions are in some cases very useful and in other cases are completely inappropriate. This ambiguity arises because there are several important "size effects" involved in fracture. If we make first the assumption that K_c is a material property and we decide to test specimens and full-scale structures which are geometrically similar, including the cracks they contain, then the elastic solutions allow us to relate the stresses σ at fracture to the crack lengths a.

In general,

$$\frac{\sigma_{\text{specimen}}}{\sigma_{\text{structure}}} = \left(\frac{a_{\text{specimen}}}{a_{\text{structure}}}\right)^{1/2} = (\text{scale ratio})^{1/2}$$

Hence, if the structure is to be used at nominal stress levels of say 5/8 of yield or greater, it will be impossible to test a small-scale model without exceeding the yield stress. We conclude that test specimens must contain proportionally larger cracks

C.K.H. Dharan et al., *Finnie's Notes on Fracture Mechanics*, DOI 10.1007/978-1-4939-2477-6_5

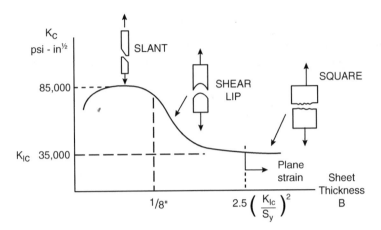

Fig. 5.1 Fracture toughness of 7075-T6 aluminum

than the structure they simulate and that the effect of plastic deformation in a test specimen will probably be more important than that in a large structure. As a result, the procedures of linear elastic fracture mechanics have been used with success to study fracture in large structures made of relatively low-strength materials such as bridges, ships, and aircraft pressure cabins. As we will see, a major difficulty in studying these cases may be that of obtaining fracture toughness data from specimens of modest size.

Another effect of size appears when measurements of K_c are made on specimens of varying thickness. A typical result is shown in Fig. 5.1 and shows that one has to be very careful in testing thin specimens to predict the behavior of thick plates. In some materials, metallurgical variables may be involved when plates of different thickness are being tested. However, for results such as shown in Fig. 5.1, it appears that the decrease in K_c with increasing thickness is due primarily to the development of triaxial stresses in the crack tip region as plate thickness is increased. This plane stress to plane strain transition results in a great decrease in the extent of the plastic zone relative to the plate thickness which is reflected in a decrease in the fracture toughness. Since the size of the plastically deformed region relative to the plate thickness plays such a major role in fracture studies, we will discuss methods for estimating its extent in some detail. Also, as the presence of the plastic region at the crack tip perturbs the elastic stress field in this region, we will discuss methods of modifying the elastic solution to allow for this factor.

Referring again to Fig. 5.1, the general appearance of the fracture surface after the crack has propagated for some distance is sketched for specimens of three thicknesses. Although such behavior is typical of steels, aluminum alloys, and titanium alloys, one has to be careful in identifying fracture appearance with a position on the curve of K_c versus thickness. Beryllium, for example, shows a variation of K_c with thickness, but the fracture surface is square on a macroscopic scale even in very thin sheets. Side grooving, which is often employed in double cantilever beam specimens, may suppress the shear lip and lead to the conclusion that plane strain had been achieved even though this is not the case.

We are referring in Fig. 5.1 to specimens such as that shown in Fig. 4.15 which are normally Mode I loading. However, under plane stress conditions, once the crack has propagated out of the starter cut or crack, a slant fracture will generally develop. Under these conditions, the crack front is subjected to a mixed Mode I and Mode III loading. For this reason, the notation K_{IC} is reserved for the plane strain fracture toughness with other values being referred to merely as K_c. It is important for consistency among different investigators that the limits for K_{IC} measurement be defined. For the specimens shown in Figs. 4.14 and 4.15, ASTM Standard E-399 [2] sets the value shown in Fig. 5.1 of $B \geq 2.5 \left(K_{Ic} / S_y \right)^2$. We will discuss the basis of this value later in the chapter. At first sight, this rule would appear to preclude plane strain testing of lower yield strength steels except in extremely large thicknesses. However, lower temperatures and higher loading rates both increase the yield and decrease the fracture toughness of low-strength steels. Under these conditions, the linear elastic solutions have proven useful for studying fracture using specimens of modest size.

5.2 Estimates of Plastic Zone Size and Correction Factors

Various approximate methods have been used to estimate the size of the plastic zone. Exact computer solutions and certain analytical solutions are available for special cases, but simple approximate approaches are often adequate for design purposes. First, let us look at the elastic solutions for the stresses near the tip of a crack and determine where they exceed the yield condition. While not an elastic–plastic solution, this simple approach should give an approximate idea of the size of the plastic zone. Taking the effective stress yield criterion, yield occurs when

$$\frac{1}{\sqrt{2}}\left\{ (\sigma_x - \sigma_y)^2 + (\sigma_y - \sigma_z)^2 + (\sigma_z - \sigma_x)^2 + 6\left(\tau_{xy}^2 + \tau_{yz}^2 + \tau_{zx}^2 \right) \right\}^{1/2} \geq S_y \quad (5.1)$$

For plane stress, using Eq. (3.1) on page 66, Ch. 3, for the stress components, the preceding equation becomes

$$\left(\frac{K_1}{\sqrt{\pi r}} \right)^2 \frac{1}{2} \cos^2 \frac{\theta}{2} \left(1 + 3 \sin^2 \frac{\theta}{2} \right) \geq S_y^2$$

or

$$r \leq \frac{1}{2} \cos^2 \frac{\theta}{2} \left(1 + 3 \sin^2 \frac{\theta}{2} \right) \left(\frac{K_1}{\sqrt{\pi} S_y} \right)^2 \quad (5.2)$$

Fig. 5.2 Region in which
elastic stresses exceed yield
for Mode I crack

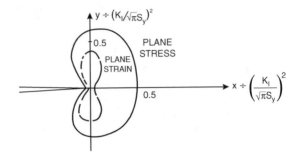

While for plane strain with $\varepsilon_z = 0$ and $\sigma_z = v(\sigma_x + \sigma_y)$, we obtain

$$r \le \frac{1}{2}\cos^2\frac{\theta}{2}\left((1-2v)^2 + 3\sin^2\frac{\theta}{2}\right)\left(\frac{K_1}{\sqrt{\pi S_y}}\right)^2 \tag{5.3}$$

We sketch these results in Fig. 5.2 to show the regions in which the elastic stresses exceed yield. The coordinates x and y have been made dimensionless by dividing it by $\left(K_1/\sqrt{\pi S_y}\right)^2$.

We could also use the maximum shear criterion for predicting yield. Knowing σ_x, σ_y, τ_{xy}, we can use Mohr's circle construction to find the principal stresses σ_1, σ_2 in the x, y plane $(\sigma_1 \ge \sigma_2)$.

$$\sigma_1, \sigma_2 = \frac{K_1}{(2\pi r)^{1/2}}\left[\cos\frac{\theta}{2} \pm \cos\frac{\theta}{2}\sin\frac{\theta}{2}\right]$$

Hence, the maximum shear stress in the x, y plane is

$$(\tau_{\max})_{xy} = \frac{K_1}{(2\pi r)^{1/2}}\cos\frac{\theta}{2}\sin\frac{\theta}{2} = \frac{K_1}{(2\pi r)^{1/2}}\frac{\sin\theta}{2} \tag{5.4}$$

This shear stress could be measured directly by testing a thin sheet of a photoelastic material containing a crack or by applying a photoelastic coating to a thin metallic plate containing a crack. The "isochromatic" fringes correspond to constant values of $(\tau_{\max})_{xy}$, that is, to $\sin\theta/r^{1/2}$ (constant), where 1, 2, 3, etc., and the constant depend on the stress intensity factor and the sensitivity of the photoelastic coating. These fringes are sketched in Fig. 5.3. The observation of such patterns is a useful method for measuring the stress intensity factor for if $(\tau_{\max})_{xy}$ is plotted against $r^{-1/2}$, a straight line of slope $K_1/[2(2\pi)^{1/2}]$ should result, at least near the crack. However, if we consider thin sheets in plane stress where $\sigma_z = \sigma_3 = 0$, the maximum shear stress is not $(\sigma_1 - \sigma_2)/2$ but $(\sigma_1 - \sigma_3)/2 = \sigma_1/2$. Hence, to predict the region in which elastic stresses exceed yield, we set $\sigma_1/2 = S_y/2$ to obtain

Fig. 5.3 Schematic of
fringe lines at tip of crack
for Mode I loading

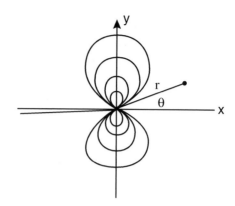

$$\frac{K_1}{(2\pi r)^{1/2}}\cos\frac{\theta}{2}\left(1+\sin\frac{\theta}{2}\right)\ge S_y$$

$$r\le\frac{1}{2}\cos^2\frac{\theta}{2}\left(1+\sin\frac{\theta}{2}\right)\left(\frac{K_1}{\sqrt{\pi}S_y}\right)^2$$

which can be compared with Eq. (5.2). For plane strain, $\varepsilon_z = 0$. So

$$\sigma_z = \nu\left(\sigma_x+\sigma_y\right) = \frac{K_1}{(2\pi r)^{1/2}}2\nu\cos\frac{\theta}{2}$$

Now, to determine if the maximum shear stress is $(\sigma_1 - \sigma_3)/2$ or $(\sigma_1 - \sigma_2)/2$, we
must examine

$$\frac{\sigma_1}{\sigma_3} = \frac{\cos\frac{\theta}{2}\left(1-\sin\frac{\theta}{2}\right)}{2\nu\cos\frac{\theta}{2}} = \frac{1-\sin\frac{\theta}{2}}{2\nu} = \frac{1-\sin\frac{\theta}{2}}{0.6}\quad(\text{for }\nu = 0.3)$$

For $0° < \theta < 48°$, $\sin\theta/2$ is small enough so that $\sigma_2/\sigma_3 > 1$. Thus, the maximum
shear stress is $(\sigma_1 - \sigma_3)/2$ in this region, and we find

$$r\le\frac{1}{2}\cos^2\frac{\theta}{2}\left[1-2\nu+\sin\frac{\theta}{2}\right]^2\left(\frac{K_1}{\sqrt{\pi}S_y}\right)^2 \qquad (5.5)$$

For $48° < \theta < 180°$, the maximum shear stress is $(\sigma_1 - \sigma_2)/2$, so in Eq. (5.4), we
put $(\tau_{max})_{xy} = S_y/2$ and obtain

$$r \le \frac{1}{2}\sin^2\theta \left(\frac{K_1}{\sqrt{\pi S_y}}\right)^2 \tag{5.6}$$

Equations (5.5) and (5.6) may be compared with Eq. (5.3), the plane strain result based on the effective stress yield criterion. Similar calculations may be made for the Mode II crack, and Fig. 5.4 shows the result, schematically, for the effective stress yield criterion.

For the Mode III crack, the only stress components are τ_{xz}, τ_{yz}. The maximum shear stress is $\left(\tau_{xz}^2 + \tau_{yz}^2\right)^{1/2}$, and from Eq. (3.3) on page 67, Ch. 3, we find

$$\tau_{\max} = \frac{K_{III}}{(2\pi r)^{1/2}}$$

Thus, as shown in Fig. 5.5, the elastic stresses exceed yield in the circular region

$$r < \frac{1}{2}\left(\frac{K_{III}}{\sqrt{\pi \tau_y}}\right)^2$$

where τ_y is the yield stress in shear. Using the effective stress yield criterion, $\tau_y = S_y/\sqrt{3}$, while $\tau_y = S_y/2$ for the maximum shear yield criterion.

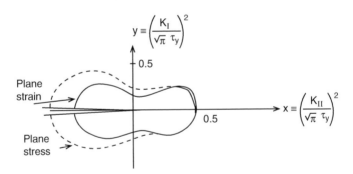

Fig. 5.4 Region in which elastic stresses exceed yield for Mode II crack

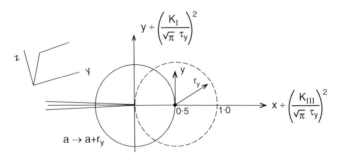

Fig. 5.5 Region in which elastic stresses exceed yield (*solid circle*) and elastic–plastic solution (*dashed circle*) for Mode III crack

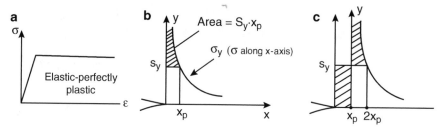

Fig. 5.6 Argument to show that plastic zone

The regions we have been discussing, in which the elastic stresses exceed yield, do not, of course, give the elastic–plastic boundary once plastic deformation has occurred. We can see this intuitively by considering the Mode I crack for the case of plane stress. Directly ahead of the crack, along the x axis, the elastic stresses are $\sigma_x = \sigma_y = K_1/(2\pi r)^{1/2}$. For equal biaxial tension, the effective stress is $\bar{\sigma} = \sigma_x = \sigma_y$. Hence, the distance ahead of the crack at which the elastic stresses exceed yield is $x_p = (1/2\pi)\,(K_1/S_y)^2$. If we consider an elastic–perfectly plastic material, as shown in Fig. 5.6a, the σ_x and σ_y stresses ahead of the crack (for plane stress) cannot exceed the yield stress S_y. Hence, if we merely truncate the elastic stress distribution at a stress $\sigma_y = S_y$, we have lost the load-carrying capacity of the region shaded in Fig. 5.6b, that is,

$$\int_0^{x_p} \frac{K_1}{(2\pi x)^{1/2}}\,dx - S_y x_p = \frac{2K_1}{(2\pi)^{1/2}} x_p^{1/2} - S_y x_p = \frac{1}{2\pi}\left(\frac{K_1}{S_y}\right)^2 S_y = x_p S_y$$

To get an elastic–plastic stress distribution with the same area under it as the original elastic stress distribution, we could move the elastic stress distribution forward by an amount x_p as shown in Fig. 5.6c. This argument is only intuitive but does suggest that for plane stress the elastic–plastic solution should show a plastic zone extending ahead of the crack by about twice the distance at which the stresses of the purely elastic solution exceed yield. One case in which a rigorous solution has been obtained is the Mode III crack in an elastic–perfectly plastic solid. It has been shown that when the plastic zone is small, it is a circular region of radius $r_y = (1/2\pi)\,(K_{III}/\tau_y)^2$ as sketched in Fig. 5.5. This plastic zone extends ahead of the crack by twice the distance ahead of the crack at which the elastic solution exceeds the yield condition. Moreover, the stresses in the elastic region can be calculated from a purely elastic solution by replacing the actual crack length a by a fictitious length $a + r_y$. This approach becomes inaccurate when the average stress on the reduced section exceeds 80 % of the yield strength [3].

Since no "exact" solution of this type has yet been obtained for Mode I (or II) cracks, it has been assumed, based on the Mode III solution and the intuitive argument of Fig. 5.6, that the plastic zone extends ahead of the crack by

Fig. 5.7 Actual crack of
length 2 (at ry)

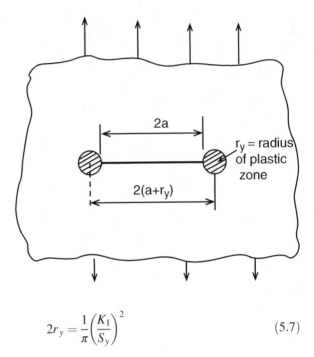

$$2r_y = \frac{1}{\pi}\left(\frac{K_1}{S_y}\right)^2 \tag{5.7}$$

It is also assumed as in the Mode III solution that the effective crack half length is
$a + r_y$ and this quantity replaces a in the equations for the stress intensity factor.
This treatment is due to Irwin [4], and he represents the plastic zone as the shaded
area shown in Fig. 5.7. The outline is left vague, deliberately, but is usually
visualized as a circle of radius r_y. This approach is, of course, empirical and its
main justification is that approximately the same fracture toughness is deduced
from plane stress fracture toughness tests using varying crack lengths.

For example, for the infinite plate under uniaxial tensile stress σ, we replace a by
$a + r_y$ in the expression for the stress intensity factor $K_1 = \sigma(\pi a)^{1/2}$, and Eq. (5.7)
leads to

$$r_y = \frac{1}{2\pi}\left(\frac{\sigma^2 \pi(a + r_y)}{S_y^2}\right) \quad \text{or} \quad \frac{2r_y}{a} = \frac{1}{\pi}\frac{(\sigma/S_y)^2}{1 - (1/2)\,(\sigma/S_y)^2} \tag{5.8}$$

Thus, the stress σ must be well below the yield stress S_y under plane stress
conditions if the plastic zone size is to be small relative to the crack length.

Irwin's physical insight in making this plastic zone correction has been confirmed not only by experiment but also by more recent analytical studies. Figure 5.8
shows results obtained by Hilton and Hutchinson [5] using a combination of
analytical solution and finite element techniques for incremental plasticity. These
authors predict the fracture initiation stress on the basis of a critical strain intensity
factor, an aspect that we will treat later. The important point for our present

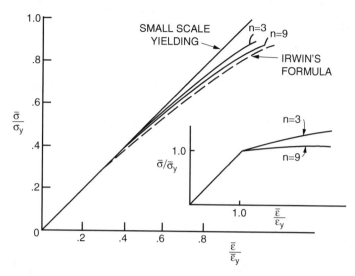

Fig. 5.8 The critical initiation stress for large-scale yielding tension

discussion is that Irwin's correction leads to fracture predictions which are very close to these more exact calculations up to stress levels of say 80 % of yield.

The Dugdale Model

Another method for estimating the plastic zone size for plane stress was given by Dugdale [6]. He considered large plates of an elastic–perfectly plastic material and assumed that the plastic zone forms essentially as an extension of the crack as shown in Fig. 5.9. From experimental observations, this appears to be a reasonable assumption for a material such as low-carbon steel. Dugdale superimposed the solutions for the two loadings shown in Fig. 5.10 to obtain the plastic zone size. Using, essentially, the approach advocated by Barenblatt and other Russian workers, he reasoned that the stress intensity factors due to the applied loading and the "crack closing forces" should be equal and opposite so that the stress is finite at the end of the plastic zone.

Taking Eq. (4.9a) and replacing a by l

$$2\left(\frac{l}{\pi}\right)^{1/2}\int_0^l \frac{\sigma}{\left(l^2 - x^2\right)^{1/2}}dx = 2\left(\frac{l}{\pi}\right)^{1/2}\int_a^l \frac{S_y}{\left(l^2 - x^2\right)^{1/2}}dx$$

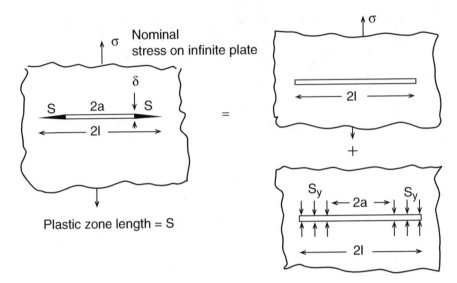

Plastic zone length = S

Fig. 5.9 The Dugdale model

Fig. 5.10 Dugdale's results
on 0.050-inch-thick steel
sheet of mechanics and
physics of solids (8, 100–
104, 1960)

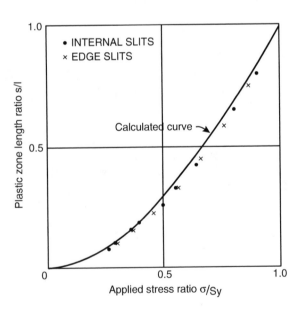

which becomes

$$\sigma \sin^{-1}\frac{x}{l}\Big|_0^l = S_y \sin^{-1}\frac{x}{l}\Big|_0^l = -S_y \cos^{-1}\frac{x}{l}\Big|_0^l$$

$$\frac{\sigma}{S_y} \cdot \frac{\pi}{2} = \cos^{-1}\frac{a}{l} \quad \text{or} \quad \frac{a}{l} = \cos\left(\frac{\sigma}{S_y} \cdot \frac{\pi}{2}\right)$$

We wish to obtain an expression for the size s of the plastic zone. Since $s = l - a$, we have

$$\frac{s}{l} = 1 - \frac{a}{l} = 1 - \cos\left(\frac{\sigma}{S_y} \cdot \frac{\pi}{2}\right) = 2\sin^2\left(\frac{\pi}{4} \frac{\sigma}{S_y}\right) \tag{5.9}$$

Experiments carried out by Dugdale, and later by others, on thin steel sheets show good agreement with experiment for nominal stresses up to about 90 % of yield (see Fig. 5.10). And it was pointed out that an edge slit may be considered to have half the length of an internal slit. Recently, the Dugdale model has been extended to treat sheets of finite size and materials which show strain hardening. Although the shape of the plastic zone in the Dugdale model is quite different from that assumed in Irwin's development, the predictions for the extent of the plastic zone ahead of the crack do not differ greatly until the nominal stress exceeds about half the yield. For the Dugdale model, we find

$$\frac{s}{a} = \frac{2\sin^2\left(\frac{\pi}{4} \frac{\sigma}{S_y}\right)}{1 - 2\sin^2\left(\frac{\pi}{4} \frac{\sigma}{S_y}\right)}$$

While for the Irwin model, $r_y = [1/(2\pi)]\ (K_1/S_y)^2$ and $K_1 = \sigma(\pi a)^{1/2}$ for the case studied by Dugdale. Making the plastic zone correction so that a becomes $a + r_y$, the value of $2r_y/a$ is given in Eq. (5.8). The values s/a and $2r_y/a$ are compared in Fig. 5.11.

Plane Strain

Because of the triaxial stress field that develops under plane strain conditions, the plastic zone size is considerably smaller than for plane stress. As a result, the elastic solutions may be applicable at higher nominal stress levels. For example, for a deeply notched circular rod, the average tensile stress on the reduced cross section can in theory be about 2.7 to 3.0 S_y before general plastic deformation begins. In practice, as pointed out by McClintock and Irwin [7], the factor is more nearly 1.6 to 1.8 perhaps due to yielding at the shoulders of the specimen. Irwin has

Fig. 5.11 Comparison of
plastic zone size for Irwin
and Dugdale models

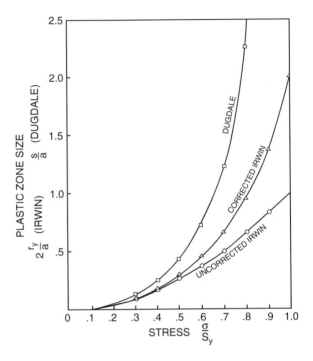

Fig. 5.11 Comparison of
plastic zone size for Irwin
and Dugdale models

suggested that a reasonable approximation to the plastic zone size in plane strain
can be obtained by replacing S_y by $\sqrt{3}S_y$ in Eq. (5.7). Thus,

$$(r_y)_{\text{plane strain}} = \frac{1}{2\pi}\left(\frac{K_{\text{Ic}}}{\sqrt{3}S_y}\right)^2 = \frac{1}{3}\frac{1}{2\pi}\left(\frac{K_{\text{Ic}}}{S_y}\right)^2 \qquad (5.10)$$

However, as the value of K_{I} for plane strain $((K_{\text{Ic}})$ is generally much less than that
for plane strain (K_c), the radius of the plastic zone for plane strain is usually very
small and is often neglected in calculating K_{I} values. We should note that whenever
the quantity r_y is referred to without additional qualifications, the plane stress value
of Eq. (5.7) is assumed. An important restriction in discussing plane strain is that
this cannot occur at a free surface where $\sigma_z = 0$ but only in regions away from the
surface as sketched in Fig. 5.12.

The ratio of the plastic zone size to the plate thickness r_y/B is an important new
dimension introduced by Irwin's development of fracture mechanics for it is this
ratio that determines whether conditions in the plastic zone, at the crack tip, will be
plane stress, plane strain, or somewhere between the two. A condition often quoted
for plane stress is $r_y \geq B$. For plane strain, the thickness condition is often given in
terms of the plastic zone size computed from the plane stress expression. That is,
using $r_y = [1/(2\pi)] (K_{\text{Ic}}/S_y)^2$ instead of say $r_y = (1/3) [1/(2\pi)] (K_{\text{Ic}}/S_y)^2$. On

Fig. 5.12 Three-
dimensional view of plastic
zone based on Irwin model

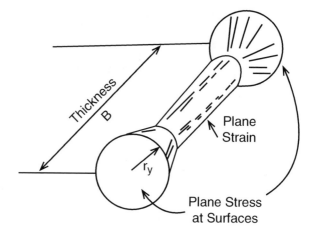

this basis, values such as $r_y/B \leq 1/10$ were originally quoted for plane strain. However, in recent years, there has been a tendency to be more conservative and set the limit for plane strain at even smaller values of (plastic zone size $\div$ plate thickness). If we are going to predict failure on the basis of an elastic stress intensity factor, the plastic zone size must be small compared to the region in which the stress intensity term provides an accurate description of the elastic stress field. Only then can we regard the fracture processes involved in the plastic zone as being controlled by the singularity term in the series expansion for the stresses. Figure 5.13 from reference [8] shows, for several conventional fracture test specimens, the error produced in the elastic stress calculations by considering only the stress intensity factor (singularity) term. Clearly, unless the plastic zone size is small compared to the crack length, we would not expect to measure the same fracture toughness values from different types of specimens. ASTM Standard Method of Test E-399-72 specifies $a \geq 2.5 \left(K_{Ic}/S_y\right)^2$ and if we take $\left(r_y\right)_{\text{plane strain}} = (1/3) \, [1/(2\pi)]$ $\left(K_{Ic}/S_y\right)^2$, this implies $\left(r_y\right)_{\text{plane strain}} \cong 0.02$. From Fig. 5.13, all the values for this value of describe the stress field with maximum errors of -2 to $+6$ %. An additional specification, mentioned earlier, which is designed to ensure plane strain is $B \geq 2.5 \left(K_{Ic}/S_y\right)^2$ which implied $\left(r_y\right)_{\text{plane strain}} \leq B/50$. In view of the restrictions which are imposed to ensure valid plane strain results, one might question the accuracy of many of the plane stress results reported in the literature since here the plastic zone sizes are relatively larger. Fortuitously, perhaps many of these tests have been conducted on large center cracked plates. The singularity terms give a reasonable approximation to the true stress field at quite large distances from the crack for this cases. In addition, the Irwin correction of replacing a by $a + r_y$ in computing stress intensity factors increases the stress levels corresponding to the singularity term. As we will see, the case of plane stress involves other aspects

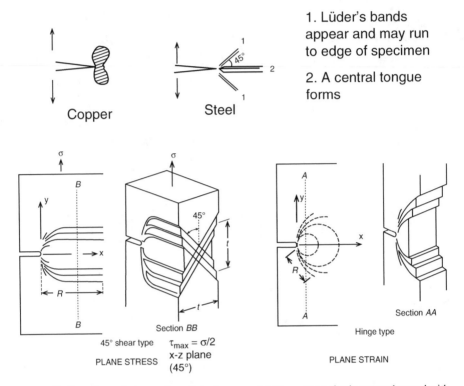

Fig. 5.13 Experimental observation of plastic zones. (**a**) Plane stress plastic zones observed with photoelastic coatings on the surface of the metal. (**b**) Plastic zones observed for plane stress and plane strain by etching silicon–iron alloys (Hahn and Rosenfield Acta Met *13*, 293, (1965)

such as slow crack growth and slant (mixed Modes I and III) fracture. However, the moral to be drawn is that the first term in the series expansion for the stresses (the singular or stress intensity term) provides only an approximation and not an exact description of the stresses away from the crack tip. For this reason, small but consistent deviations in fracture toughness values would be expected from different types of specimens.

5.3   Experimental Evidence of Plastic Zones

Because of the importance of the plastic zone size, a number of more detailed elastic–plastic analyses and computer solutions have been made which we will discuss later. Plastic zones for plane stress have been studied using photoelastic coatings and with techniques based on measuring transverse contraction near the tip of the crack. Both static and dynamic measurements are possible. For plane strain, and also for plane stress, Hahn and Rosenfield [9] have tested an iron—3 % silicon alloy which may be etched to reveal plastic zones. The deformation patterns deduced from their tests are shown in Fig. 5.13b with a showing typical photoelastic observations.

For plane stress, if $\sigma_y > \sigma_x$ in the plastic zone ahead of the crack tip, we would expect shear to occur on the two sets of planes shown in Fig. 5.13b although the manner in which this would occur at the crack tip is not clear. It is said that when $S \cong B/2$ in the Dugdale model of Fig. 5.9, that 45° shear bands are visible, and that when $S \cong 2B$, the pattern of deformation is in very good agreement with that sketched in Fig. 5.13b for plane stress. For plane strain, the transverse stress σ_z is the average of the principal stresses in the x–y plane. Thus, we would expect the planes of maximum shear stress, along which plastic deformation should be concentrated, to lie at 45° to the directions of the principal stresses in the x–y plane. Generally, the experimental observations confirm this qualitative picture.

5.4 The Plane Stress–Plane Strain Transition

Since the plastic zone size is a function of K_c/S_y and since K_c is sensitive to the ratio of the plastic zone size to the plate thickness, there is a tendency for the transition from plane stress to plane strain, or the reverse, to be self-stimulating. For example, if the plate thickness is increased, "constraint" on the plastic zone is increased, conditions become more nearly plane strain, and the ratio r_y/B decreases due to changes in r_y and B. If the size of the plastic zone decreases, the energy absorbed in fracture is less and the fracture toughness K_c decreases. A decrease in K_c in turn decreases the size of the plastic zone through $r_y = [1/(2\pi)] \left(K_c/S_y\right)^2$. The argument is qualitative, but we might expect that a small change in thickness would produce a large change in fracture toughness. Or, to say the same thing, the transition from plane stress to plane strain may occur over a small range of plate thicknesses.

This "self-stimulating" nature of the plane stress–plane strain transition allows us to explain a number of aspects of the ductile–brittle transition in steels and other materials. Many materials show an increase in yield stress as the temperature is lowered or as the rate of strain is increased. This is particularly the case with low-carbon "mild steels," the traditional steels of construction. In such a material, if a crack starts to propagate, the high strain rates near the tip of the crack will elevate the yield stress, the plastic zone becomes less, and with the transition from plane stress to plane strain, the fracture toughness decreases abruptly. As the crack speeds up, the fracture surface will show rapidly decreasing amounts of "shear lip," and so a flat fracture surface may arise as the consequence of a running crack in these materials. By contrast, the yield strength of high-strength steels is not so sensitive to strain rate, and they do not show such a pronounced transition in fracture behavior as the low-strength steels. As we will see in the next chapter, the requirement for plane strain fracture toughness tests $B \geq 2.5\left(K_{Ic}/S_y\right)^2$ cannot be satisfied for low-

strength steels with reasonable specimen sizes unless impact load, low temperatures, or both are used. However, the elevation of S_y that occurs due to the high
strain rates at a running crack may make it possible to apply linear elastic fracture
mechanics in making measurements on running cracks and crack arrest
calculations.

Although the different amounts of plastic deformation that occur around
cracks in plane stress and plane strain and the influence of strain rate and
temperature on S_y (and hence r_y) explain many aspects of the transition from
"ductile" to "brittle" fracture, this is not the whole picture and metallurgical
factors may be involved.

Quite apart from the plane stress–plane strain explanation of the transition
temperature, low- and medium-strength steels have been observed to show a
pronounced increase in K_{Ic}, the plane strain fracture toughness, with increasing
temperature. Fine grain size is desirable to get a low transition temperature, and it is
harder to produce fine grain size in large sections. So the increased transition
temperature found in heavy sections could be due to this effect as well as the
increased plane strain constraint.

Tempered martensite is a favorable microstructure for alloy steels, but it is also
hard to achieve in thick sections. For an AISI 4340 steel, the transition temperature
(Charpy V-notch 25 ft. lbs) for tempered martensite has been reported to be 350°F
below that of pearlite of the same hardness.

5.5 Elliptical or "Thumbnail" Cracks

The cracks we have been discussing are "through cracks," and the application of the
equations describing their behavior to design is fairly straightforward. However,
many of the cracks found in practice arise from such things as weld defects, arc
burns, corrosion pits, or other defects in the manufacture. These cracks are better
approximated by the thumbnail crack of Fig. 5.14. Even in thin plates, the leading
edge of such a thumbnail crack should see, essentially, plane strain constraint so the
corresponding fracture toughness may be low. As load is increased, the thumbnail will
"click" through and become a through crack. The fracture toughness K_c for a sheet of
thickness B will then determine whether the crack arrest continues to propagate.

The stress intensity factor for a surface semielliptical crack or a buried elliptical
crack may be estimated from equations given by Irwin. These were given in Chap. 4
and are summarized in Fig. 5.15.

For shallow flaws and a nominal stress well below yield, the critical flaw depth is
given by

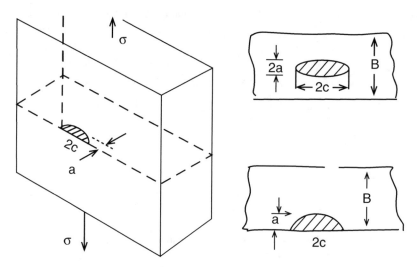

Fig. 5.14 Part through "thumbnail" crack

Surface Flaw $(a/Q)_{SF}$

Embedded Flaw $(a/Q)_{EF}$

Q : Correction factor

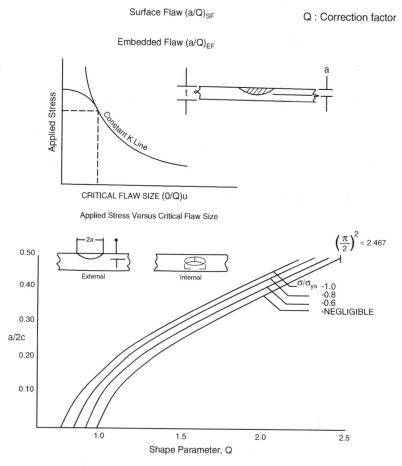

CRITICAL FLAW SIZE $(\sigma/Q)u$

Applied Stress Versus Critical Flaw Size

Fig. 5.15 Flaw shape parameter curves for surface and embedded cracks

$$a = \frac{1}{1.21\pi}\left(\frac{K_{\mathrm{Ic}}}{\sigma}\right)^2 \quad \text{for a surface flaw}$$

$$a = \frac{1}{\pi}\left(\frac{K_{\mathrm{Ic}}}{\sigma}\right)^2 \quad \text{for a buried flaw}$$

In the more general case,

    a is replaced by a/Q

where Q can be evaluated from Fig. 5.15. Irwin's analysis is based on the elliptical flaw in an infinite solid and includes corrections for plastic strains at the crack boundary and the insertion of a free surface normal to the crack through the major axis of the ellipse. Limits have been suggested such as $a < B/2$ and crack area less than 1/10 (specimen cross-sectional area) in applying this analysis. In more recent work [10], correction factors have been obtained for the free surface of the plate opposite the crack opening (back face free surface). Once the critical stress for crack propagation has been reached, the thumbnail crack "clicks" through to become a through-crack of the type we have discussed. A value often quoted for this subsequent through-crack length is $2B$. Whether the through-crack arrests will depend on its length and the value of K_c for the plate thickness. Thus, in thin sheets, with large K_c, if a small crack is permissible, it may be less restrictive to design for containment of the subsequent through-crack than it would be to design against propagation of a thumbnail crack. However, if the structure has to contain pressure for long periods, e.g., in space, then propagation of thumbnail cracks may have to be considered. In thick plates, the value of K_{Ic} may be low enough such that once the thumbnail crack grows into a through-crack, there isn't much chance of stopping it. So design has to be based on thumbnail flaws. For example, in 1965, a large rocket motor case failed while under hydrostatic test due to an internal flaw of about 0.1″ by 1.4″. The dimensions were 260″ diameter, 0.73″ well, and 65 ft. length. The yield strength is $S_y = 250,000$ psi, and the pressure at failure is 542 psi (65 % of proof).

$$\sigma = \frac{pR}{t} = \frac{542 \times 130}{0.73} = 97,000 \quad \text{psi} \quad \frac{\sigma}{S_y} = 0.39$$

$$\frac{a}{2c} = \frac{0.05}{1.4} = 0.036 \quad \text{so} \quad Q \cong 1.0 \text{ (from Fig. 5.15)}$$

$$a = 0.5 = \frac{1}{\pi}\left(\frac{K_{\mathrm{Ic}}}{\sigma}\right)^2 = \frac{1}{\pi}\left(\frac{K_{\mathrm{Ic}}}{97,000}\right)^2$$

$$K_{\mathrm{Ic}} = (\pi \times 0.05)^{1/2}97,000 \cong 38,500 \quad \text{psi} - \text{in}^{1/2}$$

Larger values of K_{Ic} had been expected for this material. However, subsequent fracture toughness tests showed that the location of the crack tip relative to the weld was an important factor. Also the fracture tests show considerable scatter. We shall return later in Chap. 8 to the question of using test data to predict the minimum likely value of fracture toughness for a large structure.

An excellent example of the use of the elliptical flaw solution to explain failure in pressure vessels was given by Tiffany and Masters [11]. Four failures in 17-7 PH stainless steel forged pressure vessels are analyzed and shown to be in good agreement with predictions. For this material, $S_y \sim 165{,}000$ psi, and $K_{Ic} \sim 50{,}000$ psi-in$^{1/2}$. Failures at nominal stress levels of about 2/3 of yield were due to elliptical or semielliptical flaws with $a \sim 0.05$ in. This emphasizes the fact that small flaws must be detectable if high-strength materials are to be used at stress levels approaching yield.

5.6   Slow Crack Growth by Fatigue and Stress Corrosion Cracking

A factor which is very important in connection with fracture of high-strength materials is slow crack growth due to environment or cyclic loads or the combination of both. Traditionally, fatigue and environmental effects were studied with smooth specimens, or specimens containing stress concentrations, and the cycles to failure or the time to failure were recorded as a function of the nominal stress. However, now with the advent of fracture mechanics, many investigators are studying crack propagation rates due to fatigue and environmental effects.

A bewildering variety of crack propagation laws have been proposed which have the form

$$\frac{da}{dN} = f(a, \sigma_a, \sigma_m)$$

where a is crack length, N the number of cycles, σ_a the alternating stress, and σ_m the mean stress. The minimum and maximum stresses in the cycle σ_{min} and σ_{max} are given by $\sigma_{min} = \sigma_m - \sigma_a$, $\sigma_{max} = \sigma_m + \sigma_a$. Often the cycle is described by σ_a and the cycle ratio $R = \sigma_{min}/\sigma_{max}$. A great simplification was introduced by Paris and his colleagues in 1961 when they pointed out that crack growth rates could be expressed in terms of K_I which combines stress, crack length, and the specimen geometry. Corresponding to σ_{max} and σ_{min} for a given crack length and configuration, K_{max} and K_{min} may be calculated. The quantity ΔK is taken as $K_{max} - K_{min}$ unless K_{min} is negative in which case $\Delta K = K_{max} - 0$. An exception to this rule is that with nonstress relieved welded constructions with unknown mean stresses, ΔK is taken to be conservative as $K_{max} - K_{min}$ even if K_{min} is negative. Except for very small or very large values of da/dN, it is found that the role of mean K is very small compared to that of the range ΔK. Figure 5.16 due to Ritchie summarizes the experimental results. In the midrange, $\frac{da}{dN} \sim c(\Delta K)^m$, $m \geq 2$. Experiments from a wide variety of specimens were plotted in the same band in a log–log plot of da/dN

MECHANICS AND MECHANISMS OF FATIGUE CRACK GROWTH

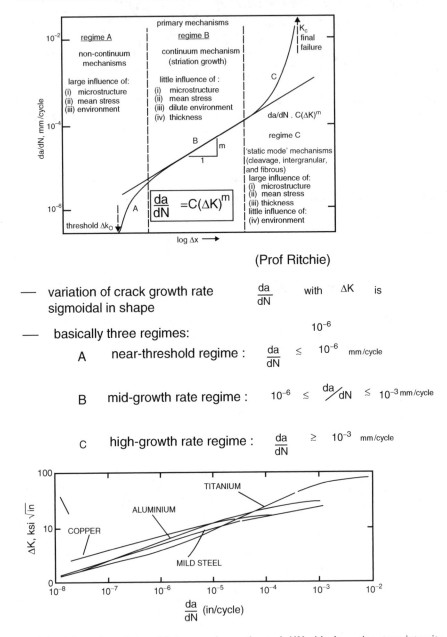

(Prof Ritchie)

— variation of crack growth rate $\frac{da}{dN}$ with ΔK is sigmoidal in shape

— basically three regimes:

      A   near-threshold regime : $\frac{da}{dN} \leq 10^{-6}$ mm/cycle

      B   mid-growth rate regime : $10^{-6} \leq \frac{da}{dN} \leq 10^{-3}$ mm/cycle

      C   high-growth rate regime : $\frac{da}{dN} \geq 10^{-3}$ mm/cycle

Fig. 5.16 Schematic variation of fatigue crack growth rate da/dN with alternating stress intensity ΔK in steels, showing regimes of primary crack growth mechanisms. Comparison of crack propagation curves, ΔK reduced to modules of aluminum

versus ΔK in this midrange. Also, tests on steels range from A36, $S_y = 36,000$ psi; from A441, $S_y = 40 \sim 50,000$ psi; and from A514, $S_y = 100,000$ psi plot within the same scatter band. By contrast to steels, organic polymers may show great sensitivity of the da/dN v. ΔK relation to changes in microstructure. In many cases, the stress levels and inherent flaw sizes in welded structures imply that life commences in Region B of Fig. 5.15.

To illustrate the type of calculation which can be made, we assume that the stress intensity factor for the flaw may be written as $K_1 = Q\sigma a^{1/2}$. The numerical factor Q (not related to the thumbnail flaw) will vary in general with flaw dimensions but here is assumed to be constant. Combining with $da/dN = c(\Delta K)^m$ leads to $da/dN = CQ^m \Delta\sigma^m a^{m/2}$. Integrating from a_i to a_c and 0 to N gives

$$\int_{a_c}^{a_i} \frac{da}{a^{(m/2)}} = CQ^m \Delta\sigma^m \int_0^N dN$$

$$N = \frac{2}{(m-2)CQ^m \Delta\sigma^m} \left[\frac{1}{a_i^{m/2-1}} - \frac{1}{a_c^{m/2-1}} \right], \quad m \neq 2.$$

For a satisfying LEFM, a_c may be determined for K_{Ic}, K_c, but since $m > 2$, it is seen that a_i is more important than a_c in determining life. Assuming $a_c \sim a_i$, $Q = $ constant, $a_i =$ constant, leads to $\Delta\sigma \sim N^{-1/m}$. Such a plot appears similar to the early part of the traditional "S–N" fatigue plot, but it has a different physical basis.

Going back to the preceding equation for N, we may establish an upper bound for a_i by proof testing at stress σ_p while a_c is determined from the operating stress σ_0 and the fracture toughness. Assuming plane strain conditions

$$K_{\text{Ic}} \geq Q\sigma_p a_i^{1/2}, \quad K_{\text{Ic}} = Q\sigma_0 a_c^{1/2}$$

Taking the equality,

$$N = \frac{2}{(m-2)\ C\Delta\sigma^m Q^2 K_{\text{Ic}}^{m-2}} \left[\sigma_p^{m-2} - \sigma_0^{m-2} \right]$$

This simple picture may be extended to more elaborate fatigue growth laws, but the concept of proof testing to guarantee a maximum initial flaw size a_i with subsequent slow growth to a flaw size a_c is basic to most analyses. The approach of plotting $\Delta\sigma - N$ curves for welded joints actually preceded fatigue crack propagation studies. The interesting point is that workers in this field found that the stress range was the important variable and all steels behaved in almost the same way. This is in striking contrast to usual S–N curves in which mean stress is important and the strength of the steel shifts the curves. Comparison of several types of joints have since shown that the slope of the S–N plots is equal to that predicted from FCP plot in Zone B. Incidentally, in the $\Delta\sigma - N$ plots for welded parts, failure is normally defined as a crack breaking through a major part of the structure so extensive crack

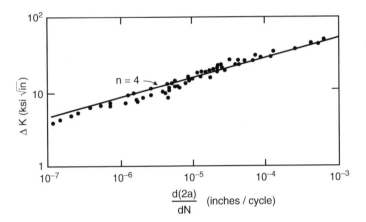

Fig. 5.17 Fatigue crack growth rate followed a fourth power law very well for 2024-T3 aluminum alloy, using through crack specimens. From Johnson and Paris, e.g., Fracture Mechanics *1*, 3–45, 1968

propagation is unresolved. By contrast, in smooth bar S–N curves, the vast majority of life is consumed in initiating the crack.

Many studies have been made of the fatigue of welded components (e.g., T. R. Gurney, Fatigue of Welded Structure, Cambridge University Press). However, reference to this work or to detailed fatigue crack propagation studies may be too elaborate for routine design. What has been done by the American Welding Society, offshore construction systems, and highway design codes is to classify welded details into a series of say five categories and to provide $\Delta\sigma - N$ curves for each of these based on extensive testing of full-size members.

Regime A in Fig. 5.16 is also of interest, but $\Delta K_{\text{threshold}}$ (often described at ΔK_0) is often too low to be of use in design. Typical values quoted for mild steels are 6.5 MN/m$^{3/2}$ for $R = 0$ and 4.3 for $R = 0.5$. A value of 1.2 MN/m$^{3/2}$ has been quoted for ΔK_0. The corresponding crack length for large structural parts is so small that we conclude life begins in Regime 3 of Fig. 5.16. There are many intriguing aspects of Regime A which will not be discussed here. These include the effects of the environment, crack closure due to residual stress, mean load, and yield strength and particularly the fact that ΔK_0 measurements from long cracks provide little information on short crack growth, that is, ΔK_0 deduced from long cracks may give very nonconservative predictions for the growth of short cracks such as those that initiate near the notches (Figs. 5.17–5.20).

5.7 Slow Growth, Pop-In, and Rate Effects

In the previous sections, we have assumed that the propagation of through cracks in a sheet of thickness B can be associated with a value of G_c (or K_c). While this approach is often satisfactory, particularly when plane strain conditions are

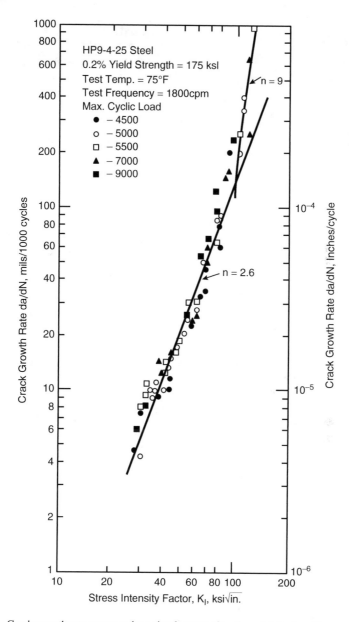

Fig. 5.18 Crack growth rate ceases to be a simple power function of ΔK when K_{Imax} approaches K_{Ic}. The good fit of the data regardless of variations in load is evidence that the stress intensity factor approach gave proper weight to both load and flaw size. From Clark Eng Fracture Mechanics *1*, 385–397, 1968

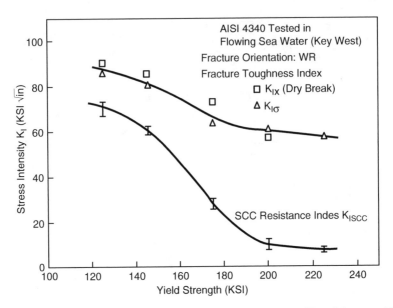

Fig. 5.19 The yield strength of 4340 steel greatly increased in seawater. From Johnson and Paris Engineering Fracture Mechanics *1*, 3–45, 1968

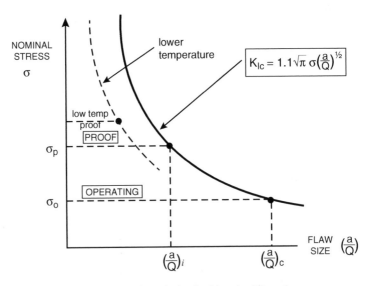

Fig. 5.20 Relation between stress and crack size for "thumbnail" crack

involved, in other cases, prolonged periods of slow crack growth under increasing load are observed. For example, Fig. 5.21 shows results obtained in testing cracked thin sheets of 2024-T3 aluminum. To describe this type of slow growth behavior, the concept of *R* curves was introduced by Irwin in 1960, which have the same units

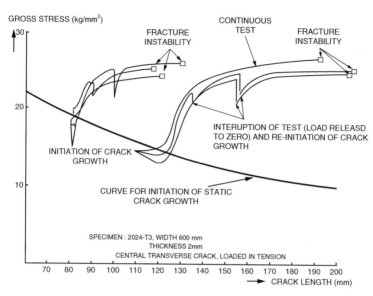

Fig. 5.21 Crack propagation under quasi-static load in continuous and interrupted tests. From D. Brock Int J. Fracture Mechanics *4* pay, 25 1968

as G, and is defined as the crack extension resistance of the material at the crack tip that opposes G. We recall that G is the strain energy release rate due to crack extension with units in lb/in^2 or lb/in. For this reason, G is sometimes referred to as the crack extension force per unit length of the crack border. Until unstable crack growth occurs $G = R$ and since at the points of unstable growth $d(G - R) = 0$, we have $(\partial G/\partial a) = (\partial R/\partial a)$ with both terms being evaluated for the value of stress at which instability occurs.

All this does not tell us very much, and the main value of the R curve concept appears to lie in the graphical presentation that one can make for slow crack growth. Consider the large plate in uniaxial tension with a central crack of length $2a_0$. Figure 5.22a shows the type of R curve that might be deduced from measurements of stress σ and crack length a (greater than a_0 due to slow growth) in thin sheets. For a different initial crack length a_0, another R curve would be obtained. In some cases, R appears to be primarily a function of crack extension and independent of initial crack length. An interesting observation is that the plastic zone around a growing crack differs from that around a stationary crack of the same length. When the plastic zone size is large, the use of elastic solutions to obtain G will be inaccurate. Relatively little information is available on this aspect of fracture—probably because of the emphasis on plane strain fracture toughness testing in the past decade. For the case of plane strain testing on thick plates, we would expect to see R curves of the type shown in Fig. 5.22b with little detectable slow crack growth preceding instability.

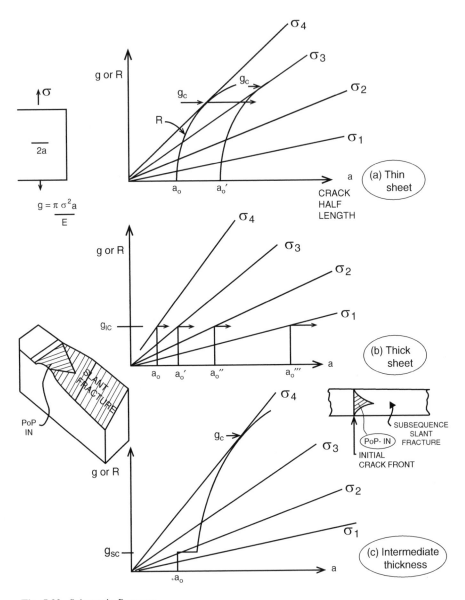

Fig. 5.22 Schematic *R* curves

For sheets of intermediate thickness, only the center region may be in plane strain. On loading, this region may "pop-in" a short distance and then arrest. This behavior and the general appearance of the fracture surface are shown in Fig. 5.22c. Such a test was used extensively in the past to estimate plane strain fracture toughness values.

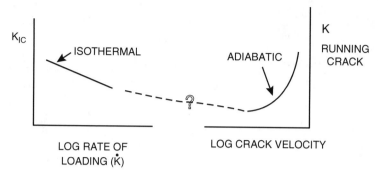

Fig. 5.23 Schematic of K_{Ic} values from crack initiation tests and K measurements on running cracks

In many cases, an R curve such as shown in Fig. 5.22c arises as a result of the transition from "square" to "slant" fracture, and once slant fracture is fully developed, the G_c value for plates of a given thickness may be relatively constant in which case it could be considered a material property. This is seen, for example, in 7075-T6 aluminum. However, Fig. 5.21 shows a situation where G_c values, obtained from different sized cracks, would differ greatly.

Another topic which we should discuss before turning to test techniques and experimental data is the effect of strain rate on fracture toughness. As sketched in Fig. 5.23, there are really two situations to consider. One is the influence of loading rate on the K_{Ic} value deduced from initiation measurements on a stationary crack. The other is the effect of crack velocity on K measurements on moving cracks. To compare these two types of tests, we could examine the rate at which stress is increasing at the edge of the plastic zone, at the time that fracture occurs.

Considering a stationary crack, subjected to a uniform external loading rate, we start by writing the stress σ_y for $\theta = 0$ and Mode I loading as

$$\sigma_y = \frac{K}{(2\pi x)^{1/2}}$$

With the plastic zone correction, this becomes

$$\sigma_y = \frac{K}{(2\pi x')^{1/2}} \quad \text{where} \quad x' = x - r_y$$

Differentiating with respect to time,

$$\frac{d\sigma_y}{dt} = \sigma_y \left(\frac{dK/dt}{K} - \frac{1}{2x'} \frac{dx'}{dt} \right)$$

For a stationary crack front,

$$\frac{dx'}{dt} = -\frac{dr_y}{dt}$$

Taking $r_y = 1/(2\pi)\ (K/S_y)^2$ with the understanding that we replace S_y by $\sqrt{3}S_y$ for plane strain,

$$\frac{dr_y}{dt} = 2r_y\left(\frac{1}{K}\frac{dK}{dt} - \frac{1}{S_y}\frac{dS_y}{dt}\right) \sim 2r_y\frac{1}{K}\frac{dK}{dt} \quad \text{since} \quad \frac{dS_y}{dt} < \frac{dK}{dt}$$

Hence,

$$\frac{d\sigma_y}{dt} \sim \frac{\sigma_y}{K}\left(1 + \frac{r_y}{x}\right)\frac{dK}{dt}$$

At the edge of the plastic zone $x' = r_y$ and $\sigma_y = S_y$, so

$$\frac{d\sigma_y}{dt} \sim \frac{2S_y}{K}\frac{dK}{dt} = \frac{2S_y}{t}$$

since $dK/dt = K/t$ for a uniform loading rate.

For a crack moving with velocity V at constant K, we have

$$\frac{dx}{dt} = -V - \frac{r_y}{dt} \sim -V$$

Hence, at the edge of the plastic zone, $\frac{d\sigma_y}{dt} \sim S_y\left(\frac{V}{2r_y}\right)$. From this approximate analysis, we see that to match the loading rates at the edge of the plastic zone, the static test must be conducted in a time

$$\frac{1}{t} = \frac{V}{4r_y}$$

Since V is often several thousand feet a second and r_y may be a fractional part of an inch, there is usually a very large gap (in stress-rate) between the two types of tests shown in Fig. 5.23.

5.8   Test Techniques and Experimental Data

Fracture toughness values can be obtained, in principle, for any configuration for which a stress intensity solution or compliance calibration is available. In 1958, an ASTM Committee was established to develop and describe test methods for fracture testing. Initially, the recommendations of this committee involved center or side notched panels, with ink staining being used to detect slow crack growth. Ink staining was later discarded as it introduces uncertain environmental effects.

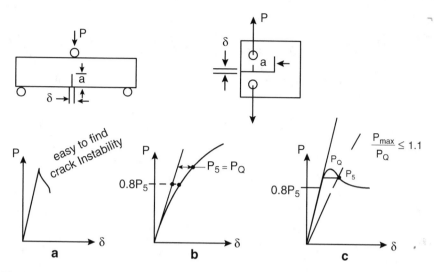

Fig. 5.24 Specimen used in ASTM-recommended test method and typical load–displacement records

Fatigue pre-cracking of machined notches was introduced as a means of producing sharp cracks. Then it became clear that fatigue pre-cracking had to be carefully controlled to avoid influencing subsequent fracture toughness values. In the mid-1960s, there was a growing emphasis on testing thick sections to obtain K_{Ic} values. The hope that "pop-in" measurements on specimens of moderate thickness would lead to reliable K_{Ic} measurement apparently did not materialize, and the need to test thick sections led to a search for specimens which could be tested at reasonable load levels. The two specimens finally selected for ASTM E399-70 "Standard Method of Test for Plane Strain Fracture Toughness of Metallic Materials"[1] were the notched bend specimen and the Manjoine or compact test specimen (also called the WOL specimen). While other tests can be used to measure K_{Ic}, the reported values do not have the "authority" of those obtained by following the ASTM test method.

The ASTM test method applies to bending or compact tension specimens with thickness 0.25 in. or greater which have been pre-cracked by fatigue under carefully prescribed conditions. The specimen sizes, loading fixture design, and displacement measurement techniques are all specified. For a valid test, the thickness B and crack length a must both exceed $2.5(K_{Ic}/S_y)^2$ where S_y is the 2/10 of 1 % offset yield strength for the temperature and loading rate of the test. Thus, until a test is carried out and K_{Ic} obtained, it is not known whether a given test will be valid.

An important part of the recommended test procedure is the interpretation of the load–displacement records. In the ideal case (Fig. 5.24a), there is no difficulty in

---

[1] In part 10 of ASTM Standards for 1982 pp. 592–692 also as an Appendix to ASTM STP 463

interpreting the record, and the load at crack instability is easily obtained. However, often, records such as those shown schematically in Fig. 5.24b or c are obtained due to slow crack growth, plastic deformation, or both of these factors. The limit on slow crack growth for valid K_{Ic} tests has been set as $\Delta a/a = 0.02$ (i.e., the maximum permissible plastic zone size). This increase in crack length corresponds to an increase in displacement (measured as shown in Fig. 5.24) of about five percent. Thus, as shown in Fig. 5.24b, the intercept P_5 on the load–displacement curve, given by a line with five percent less slope than the initial part of the data record, defines the load P_Q to be used in calculating fracture toughness. However, if as shown in Fig. 5.24c a load maximum precedes this intercept, and exceeds it, then the maximum load P_Q is used in computing fracture toughness. An obvious limitation of this procedure is that slow crack growth is not distinguished from plastic deformation. Since crack growth would be expected to produce a sharp change in the slope of the load–displacement record while plastic deformation should lead to gradual changes, the ASTM test method imposes one other requirement[2] for a valid K_{Ic} test—that the offset at 80 % of the intercept load $0.8P_s$ must be twenty-five percent or less of the offset at the intercept load P_s. The load P_Q is used to calculate a conditional fracture toughness K_Q which is taken as K_{Ic} if $a < 2.5$ $(K_Q/S_y)^2$ and $B < 2.5(K_Q/S_y)^2$.

A disadvantage of this type of testing is that the tests are expensive and time-consuming since careful pre-cracking by fatigue is required and only one data point is obtained from each specimen. In addition, it is recommended that at least three replicate tests be made.

To obtain more data from a single specimen and to study crack arrest as well as initiation, a number of investigators have used the double cantilever beam or trouser–leg specimen shown in Fig. 5.25 (and discussed earlier in Chap. 4). For this specimen, side grooving is required to control the direction of crack propagation. If deflection, rather than load, is the controlled variable, then a number of crack initiation and arrest values can usually be obtained before the crack turns out of the groove. In this type of test, fatigue pre-cracking is not often used. Once the initial machined crack has extended and arrested, the subsequent fracture toughness values should correspond to a sharp crack. Tests on 7075 T651 aluminum alloy have shown that fatigued cracks and arrested cracks gave similar fracture toughness values, but this may not be the case for all materials. The side grooves in the double cantilever specimen not only control the crack direction but appear to sharpen the point of instability in load–displacement records and promote plane strain conditions. The extent to which side grooves decrease the thickness required for valid plane strain fracture toughness measurement is not well established. This, and the fact that side grooves may promote square "flat" fracture before plane strain conditions have been reached, means that one must be cautious in reporting test

---

[2] This requirement has been replaced by a limitation on P_{max}/P_Q of +1.1.

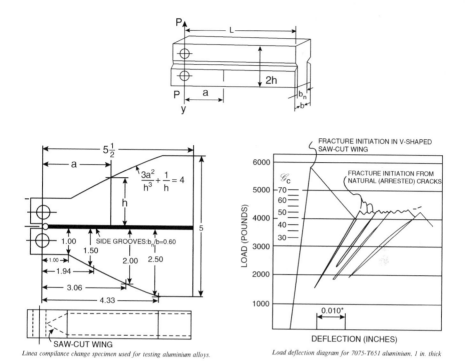

Fig. 5.25 Double cantilever beam specimen and modification used to obtain g independent of crack length (Montovey et al.)

results from double cantilever beams as K_{Ic} values. However, if the measured fracture toughness K_Q satisfies the condition

$$B_n > 2.5 \left(\frac{K_Q}{S_y}\right)^2$$

where B_n is the net thickness of the side-grooved specimen, it would be reasonable to assume that plane strain conditions have been reached.

For the specimen shown in Fig. 5.25, if each leg is treated as a cantilever of length a, the compliance is given by

$$C = \frac{24}{Eb}\int_0^a \frac{x^2}{h^3}\,dx + \frac{6(1+\nu)}{Eb}\int_0^a \frac{1}{h}\,dx$$

For constant h and $\nu = 1/3$, this becomes

$$C = \frac{2}{3EI}(a^3 + h^2 a), \quad I = \frac{1}{12}bh^3$$

However, in practice, the compliance is greater than elementary beam theory would predict because the "cantilever" of length a is not "clamped" at the end of the crack. An empirical correction for "rotation" at the built-in end of the beam has been given by Mostovey et al. (J. Materials, Vol. 2, pp. 661–681, 1967) for beams with heights from 4 to 1/2 in. They find

$$C = \frac{2}{3EI}\left((a + 0.6h)^3 + h^2 a\right)$$

By varying the thickness of the double cantilever beam (DCB) specimen as shown in Fig. 5.25 (the bat wing specimen), it is possible to obtain a specimen in which dC/da is constant along the length of the specimen. Then the fracture toughness G_c is a function only of the load as the crack traverses the specimen. This may be useful in studying stress–corrosion testing in remote environments, since only load needs to be measured to determine fracture toughness.

A great deal of information on fracture toughness is now available in the various ASTM Special Technical Publication (381, 410, 415, 463) in the Aerospace Structural Handbook, ASME technical papers, etc. One must be cautious in applying this information in design for a variety of reasons. Many of the older fracture toughness values are invalid, and many of the newer ones apply to special batches of material (e.g., vacuum melted) which may have double the fracture toughness of conventionally prepared material of the same composition and heat treatment. Fracture toughness values show scatter and are often anisotropic with quite different values being obtained when cracks are propagated in the rolling direction and across the rolling direction. In addition, fracture toughness values for welds or the heat-affected zone may differ greatly from those for the parent metal. Figure 5.26 shows results given in ASTM STP 463 (p. 237) by Brown and Srawley and some data on steels given by Steigerwald in Metal Progress, Nov. 1967. Most of the data available are for steels, titanium-based alloys, and aluminum-based alloys. Normalizing the data for K_{Ic} as a function of S_y either by density or modulus leads to a similar graph since E and ρ for these alloys are in the ratio 3:1 1/2:1. On this basis (K_{Ic}/E versus S_y/E), it has been said that the best available alloys fall along a single band. In a plot of K_{Ic} versus S_y, a straight line of slope (K_{Ic}/S_y)=constant corresponds to a given plastic zone size and hence to a minimum thickness for valid plane strain testing. For example, for a 45° line, $K_{Ic} = S_y$ and $B > 2.5(K_{Ic}/S_y)^2 = 2.5''$.

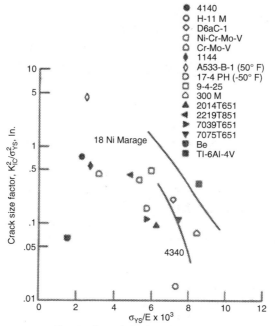

-Plane strain crack size factor for a variety of alloys in form of heavy plate or forgings. (See table I for further details).

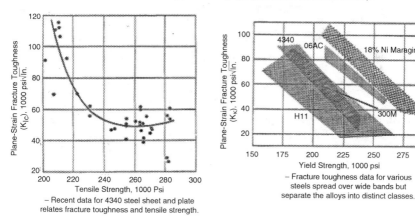

– Recent data for 4340 steel sheet and plate relates fracture toughness and tensile strength.

– Fracture toughness data for various steels spread over wide bands but separate the alloys into distinct classes.

Fig. 5.26 Typical fracture toughness

Problems

1. For the compact tension specimen, the stress intensity factor is given by

$$K = Y \frac{Pa^{1/2}}{BW}$$

where Y is a numerical quantity that depends on a/W and H/W. In testing such a specimen in a "stiff" machine, occasionally, one observes fracture initiation and arrest as sketched below.

   Assuming that the change in compliance ΔC between points 1 and 2 is entirely due to crack growth (we neglect any effect of plastic deformation on the load–displacement record), you are asked to obtain an estimate for the increase in crack length Δa between points 1 and 2. We assume that the test conditions approximate plane strain.

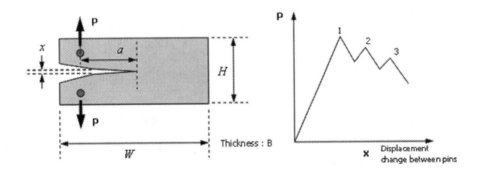

2. Griffith tested thin-walled spheres and cylinders of glass containing cracks and loaded by internal pressure to check his fracture theory. However, he was unaware of the correction factor on K that we now use to account for bulging in such configurations.

   In one of Griffith's tests, a sphere of inside diameter 1.49 in., wall thickness 0.01 in., containing a crack of length (2a) = 0.15 in., he found crack propagation to occur at a stress = 864 psi. Assuming that the maximum value of K (at any location across the wall) governs the fracture, you are asked to estimate a value of G (in-lb/in^2) from Griffith's test. (Ignore the fact that his cracks may have become blunted by stress-relieving heat treatment.)

3. We consider the situation in which a crack of length 2a (extending from $x = -a$ to $x = +a$) lies in a residual stress field which we idealize as $\sigma_y = \sigma_R \left(1 - \frac{x^2}{f^2}\right)$ for $x \ll f$. The compressive region is ignored.

   For $a \leq f$, it may be shown that the stress intensity factor at either tip of the crack due to residual stresses is $K_R = \sigma_R (\pi a)^{1/2} \left[1 - \frac{a^2}{2f^2}\right]$.

For $a > f$, the crack tip is in a compressive stress field, and K_R decreases very rapidly.

(a) If K_c is the fracture toughness of the material what is the condition, in terms of K_c, σ_R and f that cracks of any length will not propagate due to residual stresses alone.

(b) When uniaxial external loading of σ_0 psi is applied to the plate, how would you predict the critical value of a crack of length $2a < 2f$? (A one line answer is all that is needed.)

(c) Discuss, briefly, the possibility that a crack which is initiated as in part (b) will later arrest.

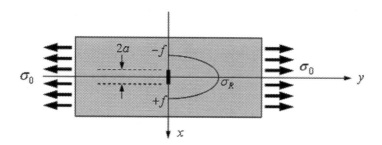

4. On July 11, 1962, on a cold winter morning, one of the spans of King's Bridge in Melbourne, Australia, collapsed. One of the main girders (a 1520 mm × 405 mm I beam) had a 125-mm-long crack in the lower flange probably due to faulty fabrication and subsequent slow crack growth by fatigue. Stress analysis has shown that the lower flange can be treated as a plate with the combined loading shown below.

(a) A uniform tensile stress of 83 MN/m2.

(b) A residual tensile stress due to welding of 260 MN/m2 over the central 100 mm of the flange.

You are asked to estimate the stress intensity factor for this combined loading. Treat the flange width as infinite in size relative to the crack. For this material, the yield strength quoted for "impact" conditions is 524 MN/m2. Values for K_c at the temperature of failure were given as
    static 139 MN/m3/2
    and dynamic (impact) 66.78 MN/m3/2.

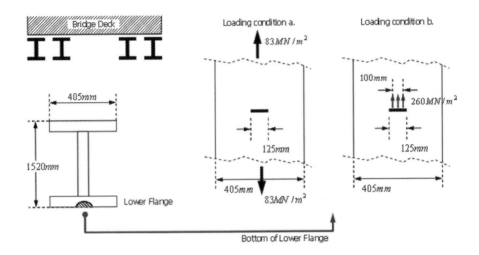

References

1. Damage Tolerant Design Handbook MCIC-HB-01, Nat. Technical Information Service, Springfield, VA.
2. Plane Strain Fracture Toughness Testing of Metallic Materials. ASTM Standard Method of Test E399-72. ASTM Standards Part 31, July 1972.
3. Koskinen MF. Trans. ASME, 85D, 585, 1963, also discussion by G.R. Irwin p. 590.
4. Irwin GR. Plastic zone near a crack and fracture toughness. p. 1V-63. In: Proc. 7th Sagamore Ordnance Materials Research Conference. Mechanical and Metallurgical Behavior of Sheet Materials, August 1960.
5. Hilton PD. Hutchinson JW. Plastic intensity factors for cracked plates, Engineering Fracture Mechanics, vol. 3, 1971, pp. 435–451.
6. Dugdale DS. Yielding of steel sheets containing slits, J Mech Phys Solids 8, 100–104 (1960)
7. McClintock FA & Irwin GR. Plasticity aspects of fracture mechanics. Fracture Toughness Testing and Its Applications, ASTM STP 1965;381:84–113.
8. Wilson WK. Contribution to discussion 'Plane strain crack toughness testing of high strength metallic materials' ASTM STP 410, 1969; 75–76.
9. Hahn GT. Rosenfield AR. Local yielding and extension of a crack under plane stress. Acta Metallurgica 1965;3:293–306.
10. Tiffany CF. Masters JN. Applied fracture mechanics. In Fracture toughness testing and its applications. ASTM STP 1965;381:249–278.
11. Tiffany CF. Masters JN. Applied fracture mechanics. In Fracture toughness testing and its applications. ASTM STP 381, 249–278 (1965).

Chapter 6
Fracture Prediction Beyond the Linear Elastic Range

6.1 Introduction

We have discussed the application of linear elastic fracture mechanics and some of its applications at length in the two preceding chapters. We recall that when the plastic zone was limited in extent, such that the full transverse constraint developed, the fracture surface was square, and material behavior could be described by a single parameter K_{Ic} (or G_c). With thinner sections and less constraint, slant fracture may develop. Under these conditions slow crack growth may occur, and then the R curve approach can be used to describe fracture behavior. One exception to this generalization is the behavior of ferritic–pearlitic steels tested below and in the transition temperature region. Here, probably because of their sensitivity to strain rate, there is an abrupt onset of rapid crack extension, and a single parameter K_c (or G_c) may still be adequate to describe fracture behavior [1]. We now turn to cases in which the assumption of small-scale yielding is inadequate and elastic–plastic or fully plastic behavior must be considered. Our emphasis again will be on cases in which full transverse constraint leads to in-plane deformation and a square fracture surface on a macroscopic level. However, a procedure analogous to R curves will be described for cases in which the fracture mode changes as the crack propagates and for tough materials which show stable tearing in plane strain.

We will consider first the traditional transition temperature test and more recent approaches to this type of measurement. As pointed out in Chap. 1, the "transition temperature" approach preceded the development of "Fracture Mechanics." For a while the two approaches to fracture prediction appeared to be quite distinct: the one based on a linear elastic analysis of stresses around cracks and applying to materials which fractured before general yield and the other applying to lower-strength steels in which fracture often occurred after general yielding—thus invalidating any elastic analysis. However, in recent years the concepts of fracture mechanics have been linked to transition temperature tests, and we will outline some of this work. Then, we will discuss more recent developments, the crack

© Springer Science+Business Media New York 2016
C.K.H. Dharan et al., *Finnie's Notes on Fracture Mechanics*,
DOI 10.1007/978-1-4939-2477-6_6

opening displacement (COD) and J integral approaches to fracture prediction. These are interrelated concepts which extend "fracture mechanics" beyond the elastic range. Very encouraging results have been obtained with these approaches which have evolved from research into general acceptance to almost the same degree as linear elastic fracture mechanics.

6.2   The Transition Temperature

For many years it has been known that in testing low-strength steels, three factors greatly enhance the ductile to brittle transition. These are triaxiality of stress (as around a notch or a crack), high strain rate, and low temperature. Thus, if cracked or notched steel parts or structures are loaded at a given rate, it is found that above a certain temperature range, the member undergoes large plastic strains before fracture occurs. Below this temperature range the member fractures abruptly before undergoing appreciable plastic deformation.

Various standard tests are used to measure transition temperatures, and usually these involve both impact loading and notched specimens. The tests can be interpreted in a variety of ways, but three general definitions are in use:

1. The temperature at which the fracture surface appearance changes. In steels the percentage of fibrous and cleavage fracture is reported.
2. The temperature at which the material's capacity for plastic deformation changes rapidly. Usually this is measured by energy absorption, but the transverse contraction at the fracture surface may also be measured.
3. The combination of stress and temperature at which a running crack is arrested.

In the past, the common tests were the Izod (with a sharp notch) or the Charpy (with a keyhole notch). Now, the most common specimen is the Charpy (with the Izod notch), which is broken in three-point bending (Fig. 6.1). Typically, the results of the tests, applied to mild steels, are shown in Fig. 6.2. The specimen is broken by a pendulum, and the energy absorbed in fracture is obtained from the difference in height of the pendulum mass, before release, and when it comes to rest after breaking the specimen. In fact, about one-third of the energy calculated in this manner may be absorbed by the testing machine, but this is usually not considered in plotting the results of the Charpy test. Among the transition temperatures used are:

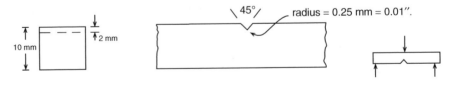

Fig. 6.1 Charpy test specimen

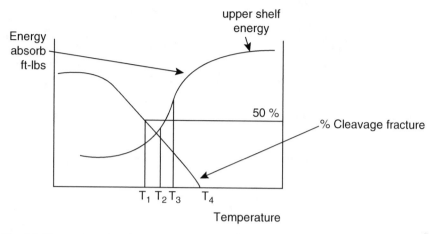

Fig. 6.2 Energy absorbed vs. temperature showing ductile-to-brittle transition (see text)

T_1: The fracture appearance transition temperature (FATT). This is quoted for a given fraction of shear (fibrous) fracture and a corresponding fraction of cleavage (flat, crystalline) fracture, e.g., 50 % cleavage.

T_2: The transition temperature based on some arbitrary level of energy absorbed, e.g., 20 J (15 ft-lbs), was often used, based on studies of Liberty Ship steels.

T_3: The midpoint of the energy curve. The rationale for this enhance is that the amount of energy absorbed in a Charpy test at a given temperature appears to have no significance unless it is interpreted in relation to the type of steel tested and the general shape and location of the energy absorption curve. For example, a recent Japanese standard specification is based on the temperature at which the Charpy V-notch energy is equal to 50 % of the "upper shelf" energy.

T_4: All fracture is in shear. This is often too conservative for economic design.

To illustrate these approaches, the following values have been quoted for a 3-½ % Ni 0.1 % C steel.

Average energy	−55 °C
$C_V = 15$ ft-lbs	−105 °C
FATT 100 % shear	−10 °C
FATT 50 %	−50 °C
FATT 70 % cleavage	−75 °C
FATT 100 % cleavage	−118 °C
Fracture initiates by tearing at notch root	−90 °C

In the Liberty ships $C_V = 15$ ft-lbs was approximately equal to FATT 70 % cleavage. In these steels, not much energy was absorbed in initiating fracture. Newer steels have lower transition temperatures and sharper breaks in the E-T curves, and $C_V = 15$ ft-lbs may correspond to 100 % cleavage. Generally, the FATT appears to be a better guide to the behavior of the structure than the energy level.

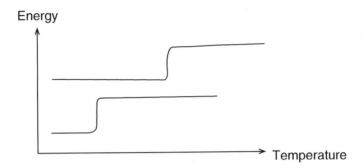

Fig. 6.3 A lower transition temperature is not necessarily a better choice between two metals with the characteristic shown

The essence then of the transition temperature approach is to be sure that the material is used above the transition temperature—as defined by a given test. Of course, attention would be given also to energy levels, and Fig. 6.3 shows a situation in which the material with the lower transition temperature might not be the better choice. An advantage of the transition temperature approach is that it does not require a detailed analysis of the stresses around cracks. Also, since the results of many tests are presented as a single number—the transition temperature—the effect of metallurgical and processing variables is easily reported. For example, various empirical relations have been devised such as

$$TT\,^{\circ}F\,(15\,ft\text{-lbs}) = 83 + 316\,C - 111\,Mn + 459\,P - 240\,Si - 8.7$$

(ASTM grain size)[1]
and presumably, the effect of irradiation could also be included in such a correlation.

As mentioned earlier, two distinct disadvantages of the transition temperature approach are that many structures are operating satisfactorily well below the transition temperature of the steel (as measured in a given test) and that some materials do not show a sharply defined transition temperature. Also, it is not clear how the transition temperature concept should be applied to parts which are thicker or thinner and have sharper or blunter notches than the specimens tested. The effects of such factors are indicated in Fig. 6.4. Despite its inadequacies, the Charpy V-notch is a simple test and still provides the vast majority of the fracture data in research studies on mild steels in addition to being a common quality control specification.

Attempts have been made to modify the Charpy test (low blow, nitrided, instrumented side grooved), and a considerable amount of analysis has been carried out on the plastic deformation in Charpy specimens prior to fracture. However, recent research work on fracture has dealt generally with specimens which contain

---

[1] Larger ASTM numbers correspond to smaller grain sizes.
   TT: Transition temperature

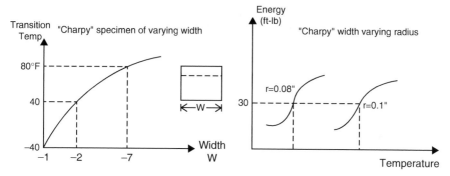

Fig. 6.4 Transition temperature v. specimen width (left). Energy absorbed v. temperature for two specimens with notch radii of 0.08″ and 0.1″

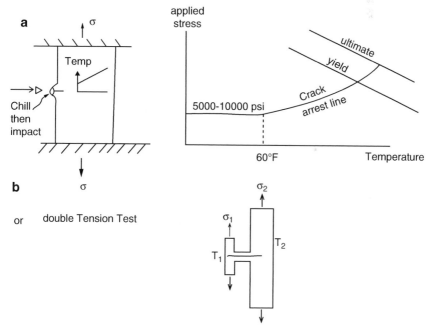

Fig. 6.5 (a) Crack arrest temperature (CAT) test configuration (b) Double tension test specimen. Applied stress v. temperature in a CAT test specimen shown the transition temperature

sharp cracks rather than machined notches (as in the Charpy specimen). Historically, one of the first tests employing sharp cracks appears to have been the Crack Arrest Temperature (CAT) test of Robertson. Typically, for a 1″ thick ship steel, the results are shown in Fig. 6.5a. The stress and temperature under which a running crack will be arrested are measured. It is a valuable test in that it specifies a safe stress, although this may to some extent depend on the nature of the loading fixture. In a thin-walled pressure vessel with pneumatic loading, the material around the crack bulges out, and the crack tip is subjected to a tearing action which may cause crack propagation at stresses below that specified by tests on flat plates. Unfortunately, the low-stress values at low temperatures imply that absolute safety (crack arrest) is, in general, not economic, and what we should worry about is the initiation of a fast running crack from a preexisting crack or other flaws rather than the arrest of a running crack. However, in special cases Crack Arrest Temperature (CAT) values may be useful in designing to limit the extent of fracture induced by accident (e.g., aircraft fuselage near propeller or fan blades, military applications, gas pipelines). As pointed out earlier in the notes, the high strain rate near the tip of a running crack raises the yield stress and hence reduces the extent of the plastic zone. For this reason it has been possible to apply the procedures of fracture mechanics to calculate K_c for running cracks and for crack arrest even in low yield strength steels.

Crack arrest tests such as the Robertson test or the double tension test shown in Fig. 6.5b are expensive to carry out. A somewhat simpler procedure is to use a bending specimen, as in Fig. 6.6a or 6.7a, and to provide a notched brittle weld bead on the tension side of the specimen. Fracture in the weld bead introduces a running crack into the specimen. Hence, at the tip of the crack, where the strain rates are high, the yield stress is elevated, and the plastic zone size is decreased relative to a corresponding static crack. A test of this type, which was referred to a great deal some years ago, is the US Navy Explosion Bulge Test (Fig. 6.6a). Tests are made at different temperatures, and the temperature above in which full plastic bulging occurs without cracking is referred to as the FTP (fracture transition plastic) temperature. The temperature above which cracks will not extend into material which is only elastically deformed, although they do extend into plastically deformed material called the FTE (fracture transition elastic) temperature. Hence at temperatures below the FTE, cracks propagate in material in which the nominal stress is below yield. Thus, in terms of the crack arrest curve shown earlier, the FTE and FTP temperatures correspond to intersection of the CAT curve with the yield and ultimate curves as shown in Fig. 6.6b. Another test is one in which a weight is dropped onto a specimen containing a notched brittle weld bead, and the deflection

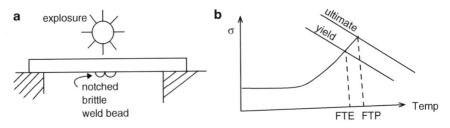

Fig. 6.6 (a) Explosion bulge test on a notched brittle weld bead (b) Definitions of FTE and FTP

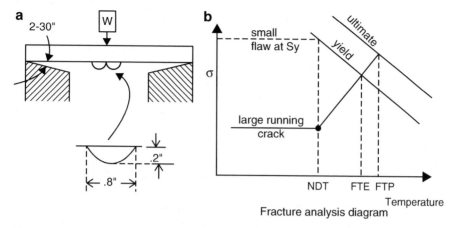

Fig. 6.7 (**a**) Drop test on specimen containing notched weld bead (**b**) NDT temperature is at knee in CAT curve

of the specimen is limited to 2–3° by stops as shown in Fig. 6.7a.[2] At high temperatures the specimen bends without cracking, while at a lower temperature, complete crack propagation occurs during the limited deformation. Thus, the NDT (nil ductility transition) temperature is defined as that at which a small (running) crack will propagate at the yield stress. This test has been widely used for quality control of structural steels, and since it appears to give a reproducible transition temperature, it has been used as the basis of design and operating procedure in certain codes.

Since both the NDT and CAT tests, if carried out on thick enough material, should correspond to plane strain fracture, we might expect the NDT temperature to correspond approximately to the temperature at which the CAT curve turns upward—if this is due to the appearance of shear fracture in the CAT test at higher temperatures. Generally, this assumption appears to be made, and the NDT temperature is shown to coincide with the knee on the CAT curve as indicated in Fig. 6.7b.

Various correlations have been presented between NDT, FTE, FTP, and the fracture appearance transition temperature (FATT) from Charpy tests such as NDT = FATT 85 % cleavage, NDT + 30° = FATT 75 % cleavage, FTE = FATT 55 % cleavage, and so on. One can only urge caution in using the various correlations, but a reasonably consistent picture for structural steels with thickness ≤ 1 and $S_y \leq 50,000\,\mathrm{psi}$ appears to be that FTE = NDT + 60°F, FTP = NDT + 120°F. Thus, design of a structure for operation at a temperature $\leq$ NDT implies that there will be no arrest of large running cracks unless the stress level is less than 5,000–100,000 psi. Also small flaws will not arrest at the yield stress. Since stress concentration due to, say, nozzle attachments in pressure vessels, or other types of discontinuities, is likely to produce local

---

[2] ASTM Standard E208-69

yielding, even when nominal stresses are low, there is thus an unacceptable high risk of "brittle" fracture if $T \leq NDT$.

At a temperature greater than NDT + 30 °F, it is found that catastrophic crack propagation will not occur for nominal stresses $\sigma \leq S_y/2$, while at temperature $\geq$ NDT + 60 °F (FTE), the arrest of even long cracks occurs if $\sigma \leq S_y/2$. For this reason $T \geq$ NDT + 60°F has been specified as a design condition in certain codes. At $T \geq$ NDT + 120°F the material behaves as if flaws were absent, and catastrophic crack propagation cannot occur at stresses below the ultimate. A large number of examples were cited by Pellini and Puzak [2] to confirm these findings. Among other aspects, they treat the role of residual stresses which have to be added to service stresses although localized residual stresses may have little influence on long cracks. As already mentioned, the "rules" are applicable for low-strength steels which have such a high upper shelf energy in the Charpy test that unstable crack propagation occurs only by cleavage with full transverse constraint. However, it has become apparent that the rules are not conservative for thicker-walled vessels, and a recent revision to Section III of the ASME code (nuclear power plant components) limits the crack arrest rule of NDT + 60 °F to sections up to 1 in. in thickness. At 6-in. thickness, the value is taken as NDT + 120 °F. Also, the "rules" may be non-conservative for large thin-walled pressure vessels with pneumatic loading. In this case, localized bulging may greatly increase the nominal stress near the crack tip.

One aspect of the design of pressure vessels for nuclear reactors which has caused concern is the elevation of the NDT temperature by neutron irradiation. The exposure is expressed as nvt (neutron/cm^2). That is, n (neutrons/cc) × velocity (cm/s) × time (s). Since an energy level above about 1 Mev is required for damage in metals, what is usually quoted is $nvt > 1$ Mev. Typically, in 20 years, present nuclear reactor pressure vessels may accumulate $1–2 \times 10^{19}$ $nvt > 1$ Mev. Figure 6.8 shows the type of results which are observed in NDT tests on specimens which have been irradiated. The importance of this result can be seen by considering

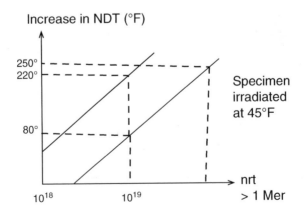

Fig. 6.8 NDT test results on irradiated specimens

current steels and present design practice. Typically for a 4″ wall thickness, the NDT temperature lies between 0 and 80 °F. Since operating temperatures are 400–600 °F, the design rule, temperature $\geq$ NDT + 60 °F, gives in the worst case a safety margin of 400-80-60 ~ 260 °F. However, this margin could be removed by irradiation. The problem becomes even more severe when start-up is considered, because the reactor has to be pressurized to come up to temperature. However, there are possibilities for annealing of radiation damage (e.g., higher temperature operation).

Since these crack arrest tests that we have been discussing give results which depend on specimen configuration and thickness, there has been a growing interest in predicting arrest by a more analytical approach (dynamic K_{Ic}, crack opening displacement). In addition, as pointed out by Nichols [3], there is a shift of opinion to prefer fracture control approaches based on prevention of initiation rather than those based on the crack arrest approach, at least for stress-relieved pressure vessels and rotating components.

Despite the limitations we have pointed out and the tendency now to focus on initiation, the Fracture Analysis Diagram of Pellini (Fig. 6.7b) was a great step forward in the transition temperature approach and allowed more rational code rules to be developed for fracture control.

We cannot leave the subject of transition temperature tests without facing a major contradiction. That is, the tests are conducted under impact conditions, yet service failures of mild steel structures have often occurred under essentially static loading. In welded structures two factors have received attention in this connection. These are "residual stresses" and "metallurgical damage." Their relative importance has been debated vigorously and at length in the literature. Additional factors such as irradiation, hydrogen embrittlement, and corrosive environments may also be involved in special cases.

Residual stresses in metals arise, in general, from nonuniform plastic deformation due to mechanical working or temperature gradients. They may also be produced by nonuniform phase changes. In welds, phase changes may occur, but the feature common to all fusion welds is the temperature gradient. The material near the weld is heated up and, being constrained by the surrounding material, it yields in compression. On subsequent cooling the material near the weld is in tension. Typically, in the weld and for a region of several times the plate thickness on either side of the weld centerline, the residual stresses have a value close to the yield stress of the annealed material. Extensive studies [4] on the fracture of welded plates have shown that in the presence of cracks, these residual stresses may lead to fracture at low nominal stress levels.

The other factor, "Metallurgical Damage," was emphasized by the work of Mylonas and coworkers at Brown University [5]. They found that by deforming a ship steel in compression to a strain level which depends on the temperature, a subsequent smooth bar tension test showed brittle behavior with fracture at nominal strains of only several percent. Some of their results are sketched in Fig. 6.9. This effect, which Mylonas refers to as exhaustion of ductility, is also observed if pre-straining is carried out in torsion [6]. The decreased pre-strain needed at about

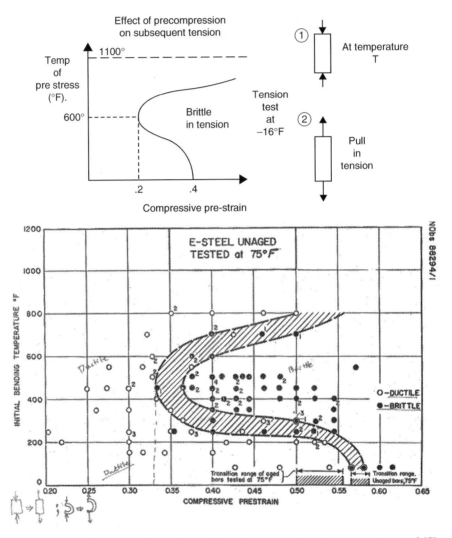

Fig. 6.9 Reversed bend tests of unaged bars of E-steel pre-strained at various temperatures (ref. [5])

600 °F suggests a strain-aging phenomenon. It has also been attributed to lines of weakness developing along boundaries of flattened grains [7]. An important observation made by Mylonas [5], and evident in Fig. 6.9, was that a stress relief anneal at 1,100 °F prevented brittle behavior in subsequent tension tests. Such a treatment also relieves residual stresses in the test specimens, and the relative importance of these two factors in producing low-stress fracture of welded mild steel structures is hard to assess. Work by Lubahn [8] attributes a greater role to residual stress than metallurgical damage.

If thermal stress relief is not possible, the proof test can redistribute stresses in a favorable manner. There is considerable evidence [9, 10] that a "warm" proof test (followed by a service temperature proof test) can greatly increase the fracture resistance of a pressure vessel. However, strain aging and large initial flaws are aspects to be considered.

6.3   The Combination of Transition Temperature and Linear Elastic Fracture Mechanics

A more quantitative approach than provided by the transition temperature is needed if we wish to predict the initiation of existing cracks. Also, as we move from low-strength to medium-strength steels, as indicated schematically in Fig. 6.10, a sharp transition temperature is no longer observed. For these reasons, there have been many attempts to extend linear elastic fracture mechanics concepts to lower-yield, higher fracture toughness materials. The basic problem is the requirement for valid K_{Ic} values that

$$B \geq 2.5 \left(\frac{K_{Ic}}{S_y}\right)^2$$

For example, for A-517-F steel which has a minimum yield of 100,000 psi, a 2-in.-thick specimen is required at $-50\,°\mathrm{F}$, and even thicker specimens are needed at higher temperatures. In this material there is a change in the microscopic fracture mode and an pronounced increase in K_{Ic} as temperature in increased [11] which is independent of the K_{Ic} to K_c transition discussed in Chap. 4.

"Compact" tension specimens of the type shown in Fig. 4.15 have been made with dimensions $24'' \times 24'' \times 10''$. However, there is an economic and practical limit to such enormous specimens. The sensitivity of yield stress to strain rate in lower-strength steels allows thinner specimens to be used in dynamic tests.

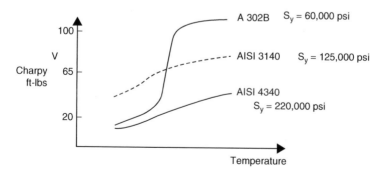

Fig. 6.10 A sharp transition temperature no longer appears when moving from low-strength to medium-strength steels

Generally, these values of K_{Id} plotted against temperature give a curve similar to K_{Ic} versus T but shifted to higher T. It has been noted [11] that the temperature shift is the same as observed in static and dynamic Charpy tests. Thus estimates of K_{Id} from K_{Ic} may be made from Charpy tests. The same type of shift has been noted to result from irradiation. One might expect even better correlation in the temperature shift between K_{Ic} and Charpy if pre-cracked Charpy specimens were used.

The observation that K_{Ic}/n (where n is the strain hardening exponent) is constant for a given steel has also been used to extrapolate K_{Ic} values to higher temperatures.

The NDT test has been interpreted in terms of fracture mechanics by Irwin [12] and by Pellini and his coworkers [13]. Irwin interpreted the markings on the fracture surface as evidence that the starter crack is a semielliptical crack of the same dimensions as the brittle weld bead ($0.2'' \times 0.8''$). Putting $\sigma = S_y$, he then obtains from the thumbnail crack solution:

$$K_{Id} = 0.78 \left(S_y\right)_{\text{dynamic}}$$

Using equations proposed for $(S_y)_d$ as a function of temperature and loading rate, he showed that K_{Id} predictions were in reasonable agreement with K values from crack arrest tests on plates. Pellini and coworkers apparently take the condition for valid plane strain tests through cracks and apply it to the thumbnail type crack produced in the NDT test. Since $B = 5/8''$ in the NDT specimen, the requirement $B \geq 2.5\left(K_{Ic}/S_y\right)^2$ or rather $B \geq 2.5\left(K_{Id}/S_{yd}\right)^2$ leads to $K_{Id} = 0.5 S_{yd}$. These workers estimate S_{yd} by arbitrarily adding 30,000 psi to the static value. The resulting estimate of the NDT temperature from values is shown to agree well with test results for several steels.[3]

A large number of empirical correlations have been presented between the results of different tests. These should only be used for the materials for which they were obtained and even then only with caution. In addition, it appears that one has to be careful in comparing crack initiation and crack arrest fracture toughness values. It has been reported that crack arrest toughness values K_{Ia} do not show the same temperature dependence as initiation values K_{Ic} in pressure vessel steels [14]. As we have seen in Chap. 4, the conditions under which a crack arrests are a function of the loading system, and test results are often difficult to interpret.

An interesting correlation between K_{Ic} and Charpy V values at 80 °F for 11 steels with yield stresses from 110,000 to 246,000 psi was reported by Rolfe, Barsom, and Gensamer [15]. The result

$$\left(\frac{K_{Ic}}{S_y}\right)^2 = \frac{5}{S_y}\left(C_V - \frac{S_y}{20}\right) \tag{6.1}$$

---

[3] More recent work has led to the estimate $K_{Id} \sim 0.6 S_{yd}$ $K_{Id} \simeq 0.6 S_{yd}$. One problem is the lack of a unique definition of S_{yd}.

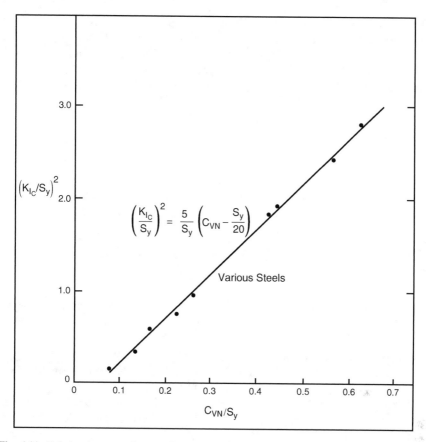

Fig. 6.11 Relation between plane strain stress-intensity factor, K_{Ic}, and Charpy V-notch energy absorption (C_{VN})

where S_y is taken as ksi and K_{Ic} as ksi-in$^{1/2}$ is shown also in Fig. 6.11. The Charpy V values are upper shelf, and all specimens showed 100 % shear. These authors use this correlation in a rather interesting way in an attempt to develop a fracture-safe design procedure. They start with the observation that there is a significant increase in through the thickness deformation when

$$B \geq \left(\frac{K_{Ic}}{S_y}\right)^2 \qquad (6.2)$$

By comparison we recall that the ASTM condition for valid plane strain tests is

$$B \geq 2.5\left(\frac{K_{Ic}}{S_y}\right)^2 \qquad (6.3)$$

The assumption made by Rolfe, Barsom, and Gensamer and also by Pellini and coworkers [16] for steels with yield stresses of 50,000 and 80,000 psi is that Eq. (6.2) guarantees sufficient through the thickness yielding so that failure will not be catastrophic. This assumption appears to need further qualification. For example, in gas pipelines, catastrophic crack propagation has occurred with slant fractures characteristic of plane stress conditions. Returning to Eq. (6.2), which is recommended only up to $B \leq 2\,\text{in}.$ in Rolfe et al. [15], the required fracture toughness is

$$K_{Ic} = S_y(B)^{1/2} \tag{6.4}$$

In view of the requirement Eq. (6.3) for valid plane strain testing, this toughness cannot be measured on plates of the thickness specified by Eq. (6.4). To avoid fracture toughness testing on thicker plates, Eq. (6.1) is combined with Eq. (6.4) to give

$$C_V = \frac{S_y[\text{ksi}]}{5}(B + 0.25), \quad S_y > 100,000\,\text{psi}, \quad B \leq 2\,\text{in}.$$

By far the most extensive correlation of different fracture tests on lower-strength steels is that provided by Pellini and coworkers. Starting with the Fracture Analysis Diagram (FAD) which we have discussed based on NDT, CAT, FTE, and FTP, they have developed drop weight tear tests[4] (DWTT) and dynamic tear tests (DT).

The dynamic tear test (DT) has been made on specimens from 5/8 in. to 12 in. in thickness. Generally, effort has been concentrated on the specimens shown which are 5/8 in. and 1 in. in thickness. Originally, DT specimens were tested by dropping weights or a swinging pendulum. However, a double pendulum machine is now commercially available for DT testing.

The advantages of the DT relative to the Charpy are said to be a sharper notch and sufficient "crack run" to develop the characteristics of a propagating fracture. By varying the dimension of the uncracked ligament, values of energy ÷ area over which the crack propagates can be plotted. Typical results are shown in Fig. 6.12 and provide information similar to R curves.

Another approach to combining the traditional transition temperature test and fracture mechanics is the use of pre-cracked Charpy specimens in an instrumented tester which allows load measurement. K_{Id} measurements are then obtained from a stress intensity solution and the observed load. Such a test has been shown [17] to give good agreement with K_{Id} values from other tests. However, an analysis [18] of the dynamic factors in the test has shown that the time at which crack propagation occurs must be determined with care if reliable K_{Id} values are to be obtained. Pre-cracked instrumented DT tests are also of interest. The possibility of replacing conventional K_{Ic} or K_{Id} tests by Charpy or DT is of considerable economic importance because of the great difference in the cost of the two types of tests.

---

[4] ASTM Standard E436-71T

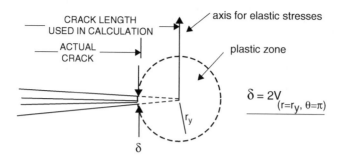

Fig. 6.12 Crack opening displacement from the Irwin model

However, there is also great interest in extending the analytical procedures of
LEFM to relate stress and critical crack size in situations where more extensive
plastic deformation occurs. Clearly, the concept of the stress intensity factor
computed from linear elastic solutions becomes meaningless in these cases. The
traditional energy approach to fracture prediction has been revived in recent years
through the development by Rice [19] of the J integral although the interpretation of
J as a fracture criterion is no longer based on the energy release rate.

An earlier approach to fracture prediction beyond the range of localized plastic
deformation is due to Wells [20]. He argues that the fracture criterion should
depend on local conditions at the crack tip. Since the fracture process depends on
the large strains in a small zone ahead of the crack, this suggested that a displacement criterion might be involved.

6.4 Crack Opening Displacement

The concept of the crack opening displacement is seen most clearly for the Dugdale
model. As we have noted this predicts the size of the plastic zone in plane stress quite
well for materials which show little strain hardening. As pointed out by Goodier and
Field [21], the Dugdale model does not predict unstable crack growth, in the sense of
Griffith's approach. As the load increases, the plastic zone size increases. If we imagine
a small increase in crack length, the plastic zone increases to absorb the energy
released. However, they suggested that fracture might occur at a critical value of the
local strain and gave an expression for the crack opening displacement (see Fig. 5.9)

$$\delta = \frac{a(8S_y)}{\pi E} \ln \sec\left(\frac{\pi}{2}\frac{\sigma}{S_y}\right)$$
$$\sim \frac{K_I^2}{ES_y} = \frac{G}{S_y}\left(\text{for } \sigma \ll S_y\right) \tag{6.5}$$

The assumption that fracture occurs at a critical value of δ has turned out to be
very useful in predicting fracture of thin-walled pressure vessels that fail in a slant

mode provided that corrections are made for outward bulging [22]. As long as the material shows the type of plastic zone postulated in the Dugdale model, the solution is valid until the nominal stress approaches yield. It is also possible to obtain a COD from the Irwin model of the plastic zone. If we imagine that crack extending to the center of the plastic zone as shown in Fig. 6.13 (in fact it will be blunted at the edge of the plastic zone), then

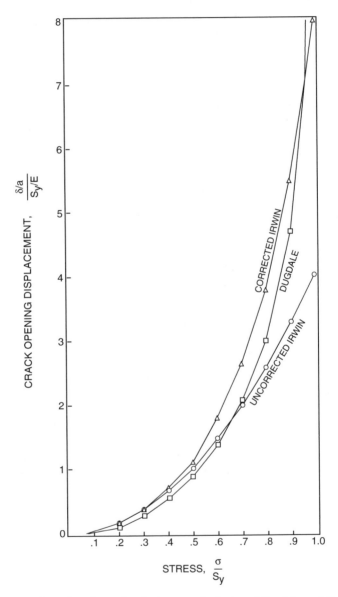

Fig. 6.13 Comparison of crack opening displacement for Irwin and Dugdale models

$$\delta = 2v_{\theta=\pi, r=r_y} = 2\frac{K_I}{G}\left(\frac{r_y}{2\pi}\right)^{1/2}\frac{2}{1+v} \qquad \text{(plane stress)}$$

$$= 2\frac{K_I}{G}\left(\frac{r_y}{2\pi}\right)^{1/2}(2-2v) \quad \text{(plane strain)}$$

Taking $r_y = (1/2\pi)(K_I/S_y)^2$ we obtain

$$\delta = \frac{4K_I^2}{2\pi S_y}\left(\frac{1}{G(1+v)}\right) = \frac{4K_I^2}{\pi S_y E} = \frac{4}{\pi}\frac{G}{S_y}$$

where, $G = \frac{K_I^2}{E}$ (plane stress)

$$\delta = \frac{2K_I^2}{2\pi S_y}\left(\frac{2-2v}{G}\right) = \frac{2K_I^2}{\pi S_y}\left(\frac{1-v}{G}\right) = \frac{4}{\pi}\frac{G}{S_y''}$$

where, $G = \frac{K_I^2}{E}(1-v^2), S_y'' = \sqrt{3}S_y$ (plane strain)

where S_y in the case of plane strain has to be taken as about $\sqrt{3}$ times the tensile yield stress (Fig. 6.13). A comparison of the COD values given by the Dugdale and Irwin (plane stress) models is shown in Fig. 6.14, and the similarity is striking.

Alternatively, one could ignore the local details of stress and strain and argue that opening up a crack by distance against a yield stress S_y in plane stress requires energy $G = \delta S_y$ per unit area. We have indicated from the Irwin plastic zone correction factor that the δ corresponding to a given value of G would have to be reduced by a factor of $\sqrt{3}$ for plane stain. More precise values have been given by elastic–plastic solutions, but before we discuss these, G has to be redefined for other than elastic behavior.

In the range of LEFM, G and K are related, and either $\delta_c, G_c,$ or K_c can be used as fracture parameters. The attraction of the COD is that it is a directly measurable quantity, at least in principle, for any type of material behavior. In practice there are obvious difficulties in measuring displacements at the crack tip, and measurements made away from the crack tip are not as easily interpreted.

A "speculative" application of the COD value is to explain the decrease in fracture toughness which occurs in thin sheets, typically [1] when $r_y > 2B$. We are dealing with plane stress and a slant fracture, so it can be argued that the COD has to be accommodated by plastic deformation over a length approximately equal to the thickness. Hence, $\varepsilon \sim \delta/B$ and since $\delta = 4K^2/E\pi S_y$

$$K = \left(\frac{E\pi S_y}{4}\varepsilon B\right)^{1/2}$$

If a critical value of the average strain is accepted as fracture criteria, then $K_c \sim B^{1/2}$ for very thin sheets. The analysis is speculative and takes no account of slow crack growth.

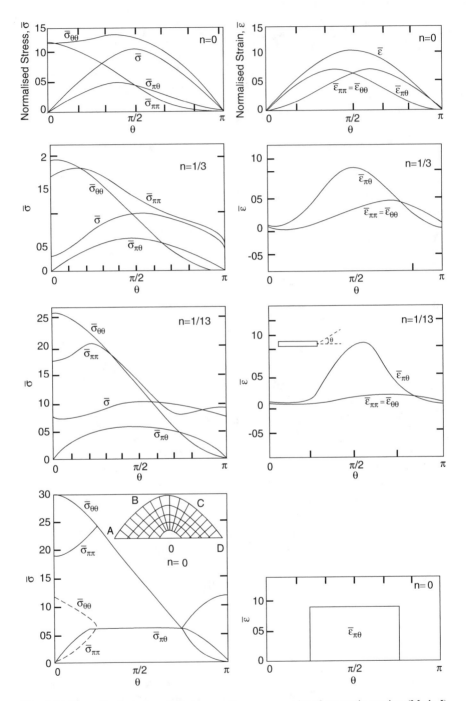

Fig. 6.14 Normalized stress and strain around a groove under plane strain tension (Mode I). For $n = 0$, solid and *dashed lines* denote double and single grooves, respectively (Hutchinson, 1968b). The Clustask–Liebowitz vol 3

COD measurements have been made primarily on bend specimens, and correlations between δ (crack tip) and opening of the crack at the surface have been presented. It has also been claimed that the ASTM K_{Ic} fatigue pre-cracking requirements are unnecessarily stringent for COD tests. Much of the COD work carried out to date has been aimed at showing that K_{Ic} values could be deduced from δ_{Ic} measurements. However, impressive progress has been made in developing design rules based on COD. The next three pages summarize these design rules.

From "The COD Design Curve" M.G. Dawes 2nd Advanced Seminar on Fracture Mechanics, Italy, April 1979, [23]

$$\bar{a}_{max} = \frac{\delta_c E \sigma_y}{2\pi\sigma_1^2} \quad \text{for} \quad \frac{\sigma_1}{\sigma_y} \leq 0.5 \tag{6.6}$$

and

$$\bar{a}_{max} = \frac{\delta_c E}{2\pi(\sigma_1 - 0.25\sigma_y)} \quad \text{for} \quad \frac{\sigma_1}{\sigma_y} \geq 0.5 \tag{6.7}$$

For general applications of Eqs. (6.6) and (6.7), the following values of σ_1 were suggested [24]:

Crack location	Weld condition	σ_1
Remote from stress concentrations	Stress relieved[a]	σ
Remote from stress concentrations	As-welded	$\sigma + \sigma_y$
Adjacent to stress concentrations	Stress relieved[a]	$\text{SCF} \times \sigma$
Adjacent to stress concentrations	As-welded	$(\text{SCF} \times \sigma) + \sigma_y$

[a]It is assumed here that the stress-relieving treatment removes all the residual stresses. In many instances this will not be so, and due allowance should be made for any stress remaining

The COD (σ_c) is measured using a three-point bend specimen and British Standard BS-5762: 1979 (methods for crack opening displacement testing values of $\bar{a}_{max}$ above are for a crack of length $2\bar{a}_{max}$ in a plate loaded by stress σ_1). For other crack shapes (e.g., thumbnail) an equivalent $\bar{a}_{max}$ may be obtained from the British document PD 6493, 1980.

The COD approach to fracture prediction, developed by the British Welding Research Institute, has been applied extensively in the design of offshore shipbuilding structures, reactive pressure vessels, and other applications. In recent years it has found increased use in the United States, for example, in pipelines.

For numerical solution it is often convenient to adopt a definition of δ as the distance shown below:

Using this definition, finite element analysis of bend specimens and cracked pressure vessels has shown remarkable correlation well beyond the range of LEFM (Welding Research Council Bulletin 299 Nov. 1984). Predictions of bars +pressure for fire vessels were within 7 % of experimental results compared to 22–44 % using COD design rule. However, extensive finite element calculations are required.

6.5   The *J* Integral

It was pointed out in Chap. 4 that the energy released by crack extension could be obtained by considering an arbitrary region around the tip of the crack, rather than taking the entire body in the manner of Griffith. This is the physical basis of the *J* integral given by Rice [19] as

$$J = \int_\Gamma W dy - \underline{T} \cdot \frac{\partial u}{\partial x} ds \tag{6.8}$$

where the quantities are defined in Chap. 4. Rice has shown that the integral is path independent and applies to nonlinear as well as linear elasticity. For the linear case it is identical to the energy release rate $G = -\partial(\text{potential energy})/\partial a$.

The *J* integral which applies to plane problems has been extended to the axisymmetric case, and other path-independent integrals have been proposed although none of these have yet found extensive application in fracture testing. An interesting modification of Eq. (6.8) for nonlinear elasticity and finite strains has been used to predict fracture in sheets of rubbery polymers using the criterion $J = J_c$.

For plastic deformation around sharp cracks, McClintock has shown that solutions obtained by Rice and Rosengren and Hutchinson for the stress and strain singularities at the crack tip may be written in terms of *J*. Assuming the power law expression $\bar{\sigma} = \bar{\sigma}_1 (\bar{\varepsilon}^p)^n$, where $\bar{\sigma}_1$ and n are constants, between effective stress and effective plastic strain, it may be shown that[5]

$$\frac{\sigma_{ij}(r,\theta)}{\bar{\sigma}_1} = \left(\frac{J}{\bar{\sigma}_1 I}\right)^{\frac{n}{n+1}} \frac{1}{r^{\frac{n}{n+1}}} \widetilde{\sigma}_{ij}(\theta)$$

$$\varepsilon_{ij}^p(r,\theta) = \left(\frac{J}{\bar{\sigma}_1 I}\right)^{\frac{1}{n+1}} \frac{1}{r^{\frac{1}{n+1}}} \widetilde{\varepsilon}_{ij}(\theta) \quad \text{Note}: \sigma\varepsilon \sim \frac{1}{r}$$

---

[5] Some authors use the notation $\varepsilon \sim \sigma^N$ where $N = 1/n$. $1 \leq N \leq \infty$ (no hardening); $0 \leq n \leq 1$ where $n = 0$ corresponds to no strain hardening.

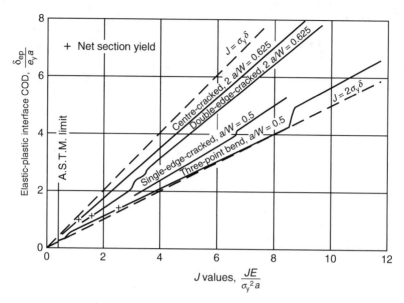

Fig. 6.15 Relationship between *J* and elastic–plastic interface COD for a range of plane strain geometries

$$u_i(r, \theta) = \left(\frac{J}{\bar{\sigma}_1 I}\right)^{\frac{n}{n+1}} r \frac{n}{n+1} \tilde{u}_i(\theta)$$

The function *I*, tabulated below, was chosen so that the maximum value of $\tilde{\sigma}_e(\theta)$ is unity

	Values of *I*			
n	1/3	1/5	1/9	1/13
Mode I Plane stress	3.86	3.41	3.03	2.87
Mode I Plane strain	5.51	5.01	4.60	4.40

Values of $\tilde{\sigma}$ and $\tilde{\varepsilon}$ are shown in Fig. 6.15.

Thus, when the singularity term dominates in a region large relative to the "pressure zone," it would be reasonable to take $J = J_{Ic}$ as a plane strain fracture criterion whether the local mechanism of failure involves stresses, plastic strains, or displacements.

A major problem in fracture research is to predict the behavior of large structures, where yielding will generally be localized, from tests on small specimens where general yield may precede fracture. Numerical studies of elastic–plastic and fully plastic stress strain fields for sharp and blunt cracks have shown *J* to be

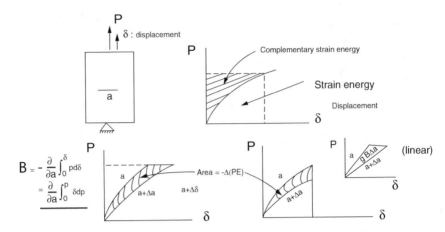

Fig. 6.16 Relationship between J and elastic–plastic interface COD for a range of plane strain geometries

approximately path independent for both deformation and incremental plasticity as long as the contour is not very close to the tip of the crack. This is not surprising since deformation theory of plasticity is equivalent to incompressible nonlinear elasticity. Also, the stress and strain distributions predicted for a monotonically loaded stationary crack from incremental plasticity would not be expected to differ greatly from those calculated using deformation theory of plasticity. Another feature which emerges from the numerical solutions is that J and the COD show a generally linear relation from the elastic range to general yield [25]. However, the constant of proportionality differs for plane stress and strain and for different specimen geometries as well as depending on the yield strain. This is illustrated in Fig. 6.16 and Tables 6.1 and 6.2.

Once general yield occurs, the interpretation of J as an energy release rate due to crack growth is no longer possible. This is because the analogy between nonlinear elasticity and plasticity no longer exists when unloading is involved, as when a crack grows from length a to $a + \Delta a$ under constant displacement. Nevertheless, a large number of studies have been devoted to the measurement of J_{Ic} from specimens which have undergone general yield. The assumption is that under certain conditions J_{Ic} values obtained on small specimens will equal G_{Ic} values obtained from specimens satisfying LEFM requirements, and ASTM test procedures for J_{Ic} measurement have been standardized. Before describing the test procedures, it may be worth offering some cautions.

First, the size of the zone (radius $\sim R$) in which the H, $R - R$, singularity dominates must be larger than the "pressure zone" in which fracture pressures occur. Generally this is of the magnitude 3–5δ. It may be larger if ahead localization seems larger and, if cleavage is involved, R has to be larger than several grain diameters.

Table 6.1 Analytical results for proportionality factor m in relationship $J = m\sigma_y\delta_t$ (plane strain)

Analytical Model, Specimen, Geometry and Degree of Plasticity	σ_Y/ε	n = 3	n = 5	n = 10	n = -	Other Hardening Values or Models	References
H, R-R singularity solution, COD at intersection of crack faces and 45° lines from tip	0.001 0.002 0.003	6.67 5.26 4.76	3.70 3.23 2.94	2.17 2.00 1.96	1.28 1.28 1.28		[29]
F.E., Deformation Theory SSy	0.002	5.00	3.13	1.92	1.43		[29]
F.E., Flow Theory, SSy	0.002	5.56	3.45	2.08	1.59		[29]
F.E., Flow Theory, SSy	0.001	-	3.33	2.13	1.54	(n = 3.33), 5.39	[8,28]
F.E., Flow Theory, SSy	0.001	-	3.70	2.13	1.51		[8,35]
F.E., Flow Theory, Finite Strains, SSy	0.003	-	3.70 3.30	2.43 2.27	1.82 1.49		[36]
F.E., CBB, Flow Theory, F.P.	0.002	5.00	-	2.04	1.64		[29]
F.E., CCP, Flow Theory, F.P.	0.002	4.17	-	1.56	1.14		[29]
F.E., CBB, Flow Theory, F.P.	-	-	-	-	-	(linear hardening) 2.00	[37,38]
F.E., Single edge notch tension, Flow Theory, F.P.	-	-	-	-	-	(linear hardening) 2.00	[37,38]
F.L., CCP, Flow Theory, F.P.	-	-	-	-	-	(linear hardening) 1.7	[37,38]
F.L., CBB, Flow Theory, F.P.					2.00		[39]

CBB = Cracted Band Bar
CCP = Center Cracted Panel

SSy = small scale yielding
F.P. = fully plastic

Table 6.2 Experimental results for proportionality factor m in relationship $J = m\sigma_y\delta_t$

Specimen Geometry Dimensions (inch)	Material	σ_y (ksi)	a	J_{lc} [lb-in/in²]	δ [inch]	m	Ref.
C88,B-0,39,w=0.39	En 32 Steel	40	4	908	0.011	2.6	[12]
C88,B-0,39,w=0.39	En 24 Steel	149	12.5	331	0.002	1.0	[12]
CCP,B-0,39,w=0.79	En 32 Steel	40	4	1016	0.011	2.2	[12]
	En 32 Steel	41	3.3	686	0.008	2.1	[40]
CCP,B-0,79,w=3.15	AP15Lx65 Pipeline Steel L-postion	74	13	1999	0.017	1.6	[41]
C88,B-0,79,w=1.57				1999	0.017	1.6	
4T-CT,B=0,side Grooves	A533-B, 93° C	61	9	914-1543	0.009-0.013	1.7-1.9	[16]
2T-CT,B=1.18; 3T-CT,B=2.95	22NiMoCr 37 Pressure Vessel Steel	77	-	-	-	1.1	[42]
2T-CT,B=1.18	22NiMoCr 37 Pressure Vessel Steel	75	-	1143	0.010	1.5	[42]
	Zircaloy E=13.9 10⁶ psi	48	13	497	0.006	1.8	[43]

CT = compact tension specimen

Second, the H, $R - R$, singularity should not be influenced by the boundaries of the specimen. This implies that some strain hardening must be present for in perfectly plastic solids, the specimen will greatly influence stresses at the crack tip. Bend and edge-cracked specimens have a triaxiality in plane strain very different from center-cracked panels.

Numerical studies suggest a limit in plane strain for the containment of the H, $R - R$, singularity in the specimen.

$B > 25J/\sigma_0$ for bend specimen $B > 200J/\sigma_0$ for center-cracked panels. Here, $\sigma_0 = (\sigma_y + \sigma_u)/2$. Comparable studies for plane stress have not been made but may be less geometry dependent.

The J integral can be interpreted as the "potential energy" difference between identically loaded bodies having a small difference in crack size, that is,

$$J = \frac{\partial (PE)}{\partial a} \quad \text{for unit thickness} \quad (6.9)$$

With this interpretation, J can be obtained by experiment much as G was obtained by compliance calibration. Although the J integral suggested and motivated this approach, we should note that Eq. (6.8) is not needed with this experimental approach. That is, we could postulate that a critical value of the rate of change of "potential energy" with crack length is the criterion for fracture and establish this value by experiment.

The potential energy per unit thickness of a two-dimensional elastic body is

$$PE = \int_A W dx dy - \int_S (T \cdot u) ds \quad (6.10)$$

where boundary tractions T are prescribed on some part of the boundary. A generalized load–displacement diagram is shown in Fig. 6.17. If boundary

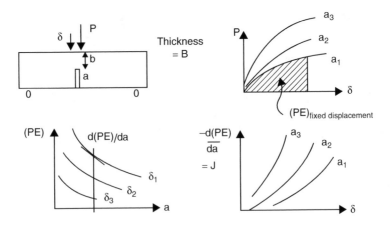

Fig. 6.17 Illustrating experimental evaluation of J from specimens with several crack lengths

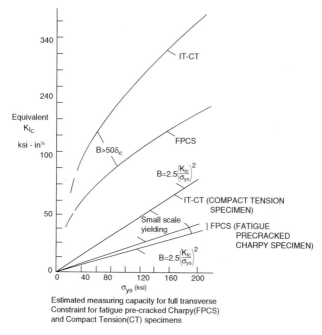

Estimated measuring capacity for full transverse
Constraint for fatigue pre-cracked Charpy(FPCS)
and Compact Tension(CT) specimens

Fig. 6.18 Estimated measuring capacity for full transverse

conditions are given in terms of force, the potential energy is equal to minus the complementary energy (the area shown shaded). When displacements are specified, the second term in Eq. (6.10) vanishes, and the potential energy is the area under the load–displacement curve. The change of potential energy with crack length is shown in Fig. 6.18 both for constant load and constant displacement. As we discussed in Chap. 4, the difference in the shaded areas is negligible if the change in crack length is small. Hence, the rate of change of potential energy with crack length is independent of the loading conditions. In practice, because the load-deflection curves become nearly horizontal after large amounts of plastic deformation, it is more convenient to evaluate the change in potential energy for a given displacement rather than a given load.

The procedure that could be followed in evaluating *J* from specimens with several crack lengths is outlined in Fig. 6.18. More convenient procedures involving tests on a single specimen [26, 27] have been proposed, and we will turn later to discuss this procedure.

An important aspect of J_{Ic} testing compared to K_{Ic} is that the cross section may be fully plastic as long as the deformation field at crack initiation is plane strain. It appears that the size limitations of linear elastic fracture mechanics may be greatly relaxed for J_{Ic} testing. Two conditions have been suggested. One is that the specimen thickness should be at least the length of the uncracked ligament. That is, $B > b$ to promote in-plane plastic deformation. The other condition is that

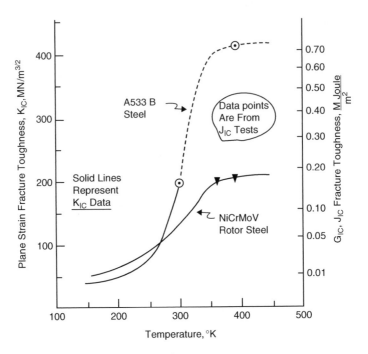

Fig. 6.19 Plane strain fracture toughness various test temperature

the thickness B should be large compared to the region of intense strain at the crack tip to ensure plane strain constraint. The crack opening displacement characterizes the regions in which critical separation processes are occurring. For plane stress, Rice has shown that $J = S_y \delta$ for the Dugdale model with no strain hardening. For plane strain, Wells [23] has quoted a reduction in δ of 2.1 for externally notched specimens and 1.7 for internal notches both in plane strain. Typically, a factor of about 2 has been obtained in other studies. It has been suggested [28] that $B \geq 25J_{Ic}/S_y$ be taken as a criterion for J_{Ic} testing, and this implies $B \geq 50\delta_{ic}$. Recall that $J = m\sigma_0 \delta_t$.

It can be seen from Fig. 6.19 that these requirements greatly increase the measurement capabilities of small specimens. Figure 6.20 shows results reported by Begley, Landes, and Wessel [29]. They indicate that fully plastic J_{Ic} tests on small specimens of a forging steel agree with G_{Ic} values obtained from an 8-in.-thick, essentially elastic compact tension specimen. Data are also shown for a 70,000 psi yield pressure vessel steel. Obviously, one has to be very careful in such comparisons to be sure that the microscopic failure mechanism does not change as a function of the type of test. However, the implication of the tests we have reported is that great progress may be possible in extending fracture mechanics to elastic–plastic and fully plastic behavior.

The measurement of J from a single test is based on the behavior of deeply notched specimens in bending. The result that will be derived is equivalent for the two cases shown in Fig. 6.21. That is, if we consider deflections due to the crack, the

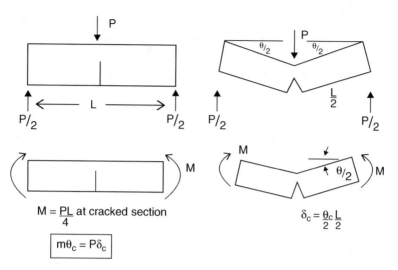

$$M = \frac{PL}{4} \text{ at cracked section}$$

$$\delta_c = \frac{\theta_c}{2}\frac{L}{2}$$

$$\boxed{m\theta_c = P\delta_c}$$

Fig. 6.20 Equivalence of $m\theta_c$ and $P\delta_c$

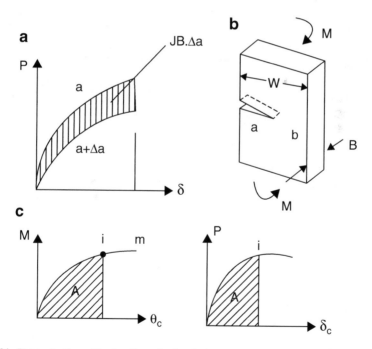

Fig. 6.21 Determination of J – bending of a deeply notched plate

case of a bar in three-point bending and one under pure bending may be treated as equivalent if crack deflections are treated as a result of solid body rotation about the cracked cross section.

From the definition of J in terms of potential energy change,

$$JB = -\frac{\partial}{\partial a}\int_0^\delta Pd\delta = \frac{\partial}{\partial a}\int_0^\delta \delta dP$$

This can be seen from Fig. 6.22a. Consider bending of the deeply notched plate shown in Fig. 6.22b.

$$JB = \int_0^M \left(\frac{\partial M}{\partial a}\right)_m dM = -\int_0^M \left(\frac{\partial \theta}{\partial b}\right)_m dM$$

Ignoring θ due to the uncracked body which is not a function of b, we saw in Chap. 4 that

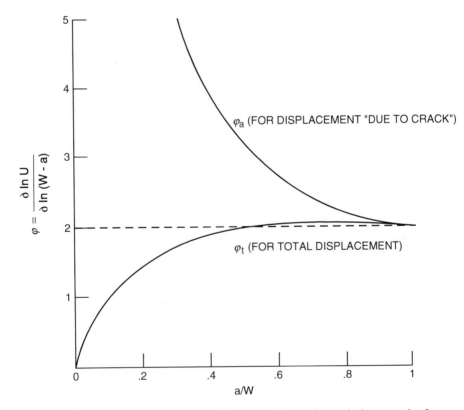

Fig. 6.22 Values of Ψ, the ratio of $J1$ to work done per unit area of uncracked cross section for an ASTM E-399 three-point band specimen of a linear elastic material

$$\theta_{\text{crack}} = \frac{15.8M}{bE'}b^2$$

It can be argued that for fully plastic or elastic–plastic behavior,

$$\theta_c = f\left(\frac{M}{Bb^2}\right)$$

In this case,

$$\left(\frac{\partial\theta_c}{\partial b}\right)_m = \frac{-2M}{Bb^3}f', \quad \left(\frac{\partial\theta_c}{\partial M}\right)_b = \frac{1}{Bb^2}f'$$

or

$$\left(\frac{\partial\theta_c}{\partial b}\right)_m = \frac{-2M}{B}\left(\frac{\partial\theta_c}{\partial M}\right)_b$$

Hence

$$J = (1/B)\int_0^M (2M/b)(\partial\theta_c/\partial M)dM = (2/bB)\int_0^\theta Md\theta_c, \quad \text{or}$$

$$J = (2A/Bb) \quad \text{[For the compact tension specimen, the numerical factor differs}$$
$$\text{from 2 and has been given as } \eta A/Bb] \quad \text{where } \eta = f(b/W)$$

$$(6.11)$$

where A is the area shown in Fig. 6.22c measured at the point of crack initiation. This remarkably simple result presents problems in application because the point of crack initiation is not easily detected. Another source of some confusion in the literature is that the result applies to deeply notched specimens with the area evaluated for the abscissa equal to the displacement (or rotation) due to the crack. Many authors have applied it to specimens with $a = W/2$ and neglected to subtract the displacement of the uncracked body. At least for linear elastic behavior, these errors are, fortuitously, self-correcting. Several authors have discussed this result. Figure 6.22 due to Srawley shows that if total displacement is used in Eq. (6.11), the result may be applied for $a/W > 0.5$, while if crack displacement is used, the expression is only valid for $a/W \geq 0.8$.

The current ASTM procedure[6] is to measure J_{Ic} using Eq. (6.11) for compact tension or bend specimens and plot J_{Ic} versus Δa the amount of crack extension—measured, e.g., by heat treating. Since a crack grows by blunting by an amount

---

[6] ASTM Standard E813-81

approximately equal to the COD/2, the value of J_{Ic} is evaluated for $\Delta a = J/2S_{flow}$ where $S_{flow} = (S_y + S_u)/2$. We note that this implies $COD = J/S_{flow}$, while other studies suggest that $COD \simeq 0.5J/S_{flow}$ may be more realistic.

Problems

1. For crack length estimates during J testing, we wish to obtain the crack length from the compliance measured during partial unloading. For an ASTM standard three-point bending specimen with span $S =$ four times the depth W and thickness B, we are given that

$$K_I = \left(\frac{3}{2}\right) B \frac{PS}{W^2} \sqrt{\pi a} f(a/W)$$

where for $a/W \ll 0.6$

$$f(a/W) = f(\alpha) = 1.09 - 1.73\alpha + 8.20\alpha^2 - 14.2\alpha^3 + 14.6\alpha^4$$

(a) You are asked to obtain an expression for the compliance due to the crack for plane strain in the form

$$(\text{compliance})_{crack} = f_1(S, B, W, E, v) f_2(a/W).$$

The numerical value of the factor should be computed for values of

$$a/W = \alpha = 0.1, \quad 0.2, \quad 0.3, \quad 0.4, \quad 0.5, \quad 0.6 \ldots$$

(b) For specimen displacement control and LEFM with $K_{Ic} = $ constant, it is said that stable crack growth is observed in the standard specimen $S/W = 4$ if $a/W \gtrsim 0.4$. Can you confirm this result? Perhaps the easiest approach is to plot K against (a/W).

(c) Repeat part (b) for the case in which the machine compliance is (1) equal to the uncracked specimen compliance and (2) twice the uncracked specimen compliance.

2. We have been asked to bid on a research project sponsored by the "National Cooperative Highway Research Program." This involves fatigue crack growth studies on steels of $S_y = 100,000 \, psi$ psi using 1-in.-thick specimens. It is specified that the loading should be zero to a maximum load such that the stress intensity factor varies from zero to 100,000 psi-in$^{1/2}$. Since our testing machine has a maximum load capacity of 15,000 lb, we have to select a specimen which allows the specified stress intensity factor to be reached at this load level. You are asked to calculate the required load for a double cantilever beam

specimen of dimensions shown below. Ignore questions of side grooving and crack stability in making this calculation.

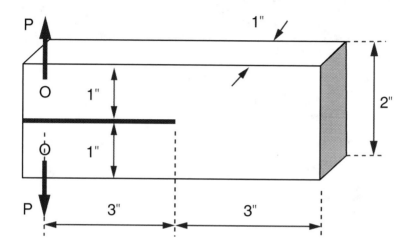

References

1. Corten HT. Testing methods for determination of fracture toughness of metals. In: Proc. 3rd int. fracture conference, vol. I, Munich; 1973.
2. Pellini WS, Puzak PP. Welding research council bulletin no. 88, May 1963.
3. Nichols RW. Some applications of fracture mechanics in power engineering. In: Proc. 3rd int. fracture, vol. 1, Munich; 1973.
4. Hall WJ, Kihara H, Soete W, Wells AA. Brittle fracture of welded plate. Englewood Cliffs, NJ: Prentice-Hall; 1967.
5. Mylonas C. Exhaustion of ductility and brittle fracture of project E-steel caused by pre-strain and aging. Dept. of Navy, Bureau of Ships Report NOBS-88294/1, Dec. 1963. See also Weld J. 36, 9-S (1957); 37, 473–5 (1958); 38, 414–5 (1959).
6. Theocaris P. The influence of torsional prestraining on notch-brittle fracture. Int J Mech Sci. 1967;9:195–204.
7. Allen NP, Early CC, Hale KF, Randall J. Iron Steel Ins. 1964;202:808.
8. Rosenstein AH, Lubahn JD. Brittle fracture of welded steel plate. Weld J. 1967;Research Supplement:481-S–90.
9. Nichols RW. Brit Weld J. Jan. 1968; 21–42; Feb. 1968; 75–84; Oct. 1968; 524–25.
10. Harrison TC, Fearnehough GD. The influence of warm prestressing on the brittle fracture of structures containing sharp defects. J. Basic Engineering 94 (2), 373–376 (1972).
11. Barsom JM, Rolfe ST. Kic transition-temperature behavior of A517-F steel. Eng Fract Mech.1971; 2:341–357.
12. Irwin GR. Linear fracture mechanics, fracture transition, and fracture control. Eng Fract Mech. 1968;1:241–257.
13. Loss FJ, Pellini WS. Practical fracture mechanics for structural steels. p. J1–35. See also many NRL reports by Pellini and co-workers.
14. Crosley PB, Ripling EJ. Crack arrest toughness of pressure vessel steels. Nucl Eng Des. 1971;17:32–45.

15. Rolfe ST, Barsom JM, Gensamer M. Paper OTC-1045, offshore technology conference, Dallas, TX; 1969.
16. Pellini WS. Analytical design procedures for metals of elastic-plastic and plastic fracture properties. In: Proc. US-Japan symposium on application of pressure vessel codes, Tokyo; March, 1973.
17. Server WL, Tetelman AS. The use of pre-cracked Charpy specimens to determine dynamic fracture toughness. Eng Fract Mech. 1972;4:367–75.
18. Turner CE, et al. Practical applications of fracture mechanics to pressure vessel technology. Inst Mech Eng (London). 1971;38–47.
19. Rice J. J App Mech. 1968;379–86.
20. Wells AA. Fracture control: past, present, future. Exp. Mech. 1973;13(10):401–410 .
21. Goodier JN, Field FA. Plastic energy dissipation in crack propagation. In: Drucker DC, Gilman JJ. Eds. Fracture of Solids, Wiley, New York. 1963; 103–118.
22. Erdogan F, Ratwani M. Fracture of cylindrical and spherical shells containing a crack. Nucl Eng Des. 1972;20:265–86.
23. Dawes, M. G., "The COD design curve," Advances in Elasto-Plastic Fracture Mechanics (2nd Advanced Seminar on Fracture Mechanics, Ispra, Italy, April 1979).
24. Dawes, M. G. (1974). Fracture control in high strength weldments. Weld. J. Res. Supp. 53, p. 369.
25. Wells AA. Fracture control: past, present, future. Exp. Mech., Vol. 13, No. 10, pp. 401–410, (1973).
26. Bucci RJ, Paris PC, Landes JD, Rice JR. J integral estimation procedures. Fracture Toughness, ASTM STP 514: 40–69 (1971).
27. Rice JR, Paris PC, Merkle JG. Some further results of J-integral analysis and estimates. In: Progress in flaw growth and fracture toughness testing. ASTM STP 536, 1973;213–245.
28. Paris PC. Discussion of ref. 24, p. 21–22.
29. Begley JA, Landes JD, Wessel ET. Third international conference on fracture, vol. IX; Munich 1973.

Chapter 7
Cleavage and Ductile Fracture Mechanisms: The Microstructural Basis of Fracture Toughness

7.1 Introduction

Our preceding discussion has dealt primarily with the macroscopic aspects of fracture with only occasional reference to behavior on the microscopic level. Eventually, it would be desirable to minimize the expensive process of fracture toughness testing by predicting behavior from more fundamental material properties and simpler material tests. Any such predictions have to be based on the microstructural features of the fracture process. Impressive progress has been made in relating fracture behavior to more fundamental properties but also because of the insight it provides into the physical processes involved in fracture.

If we exclude intergranular fracture, the two microscopic mechanisms involved in the short-time failure of metals are transgranular cleavage and ductile fracture by void nucleation, growth, and coalescence. We will not consider the ductile fracture processes which operate in fatigue to provide progressive crack extension with striation markings or the cases of failure by plastic instability, or by exceeding limit loads, which involve plastic deformation on a macroscale. As pointed out in the first chapter, the face-centered cubic metals in short-time tests at ambient temperature and in the absence of a corrosive environment will fail by a ductile fracture mode. For body-centered cubic and hexagonal close-packed metals, both cleavage and ductile fracture mechanisms may occur. In particular, the material most often studied, low carbon steel (mild steel), exhibits cleavage fracture at low temperature. When increasing temperature cleavage becomes more difficult, a rapidly increasing fraction of the fracture becomes "ductile" until at a high enough temperature ductile fracture modes dominate. Because of the attention it has received, most of our discussion of cleavage will relate to low carbon steels while the treatment of ductile fracture will be more general.

© Springer Science+Business Media New York 2016
C.K.H. Dharan et al., *Finnie's Notes on Fracture Mechanics*,
DOI 10.1007/978-1-4939-2477-6_7

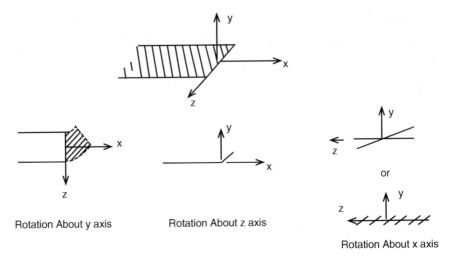

Fig. 7.1 Movement of cleavage plane entering another grain

Cleavage fracture occurs by separation along preferred crystallographic planes.[1] As a result of the flat smooth fracture surface in each grain, the surface is quite reflective and has a shiny appearance. As the cleavage crack passes from one grain to another of different orientations, it will change direction to continue propagation along the preferred crystallographic plane. Using our usual *xyz* coordinate system at the leading edge of a crack as shown in Fig. 7.1, we can describe the change in orientation between the preferred cleavage planes in adjacent grains as a rotation about the *x*, *y*, or *z* axes. Rotation about the *y* axis is the simplest case for then the crack continues to propagate in its original plane but turns as it enters the new grain. Rotation about the *z* axis, the leading edge of the crack, will cause the crack front to change direction such that the crack extension makes an angle θ in the *xy* plane with respect to its original plane. A more interesting situation is a change in grain orientation described by rotation about the *x* axis. Now, the crack cannot accommodate the new cleavage direction by turning as in the other cases. Rather the crack front may break into sections, each of which twists relative to the original crack plane to follow the new cleavage planes. A striking photograph of the breakup of the crack front in such a manner due to combined Modes I and III loading in glass has been used for many years on the cover of the journal, "Engineering Fracture Mechanics." The sections of the crack front formed at a "twist" boundary later join up and give the appearance of a river with tributaries [1]. Such changes in crack direction from grain to grain, the possibility of several cracks growing on parallel cleavage planes and forming steps where they overlap, and "cleavage tongues" which apparently form when the crack interacts with a "twinned" region [2] all lead

---

[1] Cube planes (100) in steel

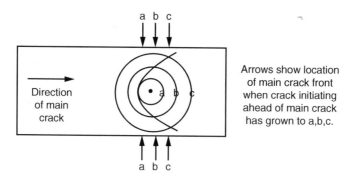

Fig. 7.2 Basis for Chevron markings

to irregularities in the fracture surface. These may be useful in failure analysis where it is often found that the local direction of crack propagation in a grain may differ considerably from the macroscopic direction. Another feature associated with cleavage fracture in mild steels on both a micro- and macroscale is a "herringbone" or chevron pattern on the surface. On a microscale this has been attributed [1, 3] to interaction of the main cleavage plane with "twinned" material. On a macroscale the chevron markings are often very pronounced. They open in the direction of crack propagation and are often very useful in tracing the growth of cracks and identifying the origin of fracture say in a pressure vessel. The mechanism of macroscopic herringbone markings may be the interaction of the main crack front with secondary fractures initiating ahead of the crack as sketched in Fig. 7.2. Following some comments on ductile fracture, we will treat very briefly some of the mechanisms proposed for cleavage fracture in steels and then discuss the cleavage fracture of notched and cracked members.

Until recently, the microstructural mechanisms involved in cleavage fracture had received considerably more attention than those involved in ductile fracture. The reason for this is that dislocation theory and the other metallurgical concepts necessary to study cleavage have been available for some time. The clarification of ductile fracture mechanisms depended on the development of continuum studies of void growth on a microscale and the details of fracture surface topography provided by the scanning electron microscope. The general features of ductile fracture are the nucleation, growth, and coalescence of voids to form a "fibrous" surface on a macroscale and a series of dimples on a microscale. A large amount of evidence indicates that the voids responsible for fracture are formed primarily at hard second-phase particles and inclusions. The nucleation of voids occurs either by fracture of particles or by decohesion of the particle-matrix interface. In some cases such as manganese sulfide particles in steel, decohesion occurs readily, while in other cases such as carbides in steel, a certain amount of plastic deformation in the matrix is needed to initiate cracking or decohesion. Voids, after being formed, grow by plastic deformation of the matrix. Their growth relative to the overall deformation of the matrix is a very strong function of the triaxiality of stress (hydrostatic stress ÷ effective stress) and the strain-hardening rate of the matrix.

Final separation results from the coalescence of the growing voids. A limit to the growth process has been taken in analytical studies as that in which the voids contact one another. In practice this is preceded by necking or localized shear between voids. These processes are also sensitive to the strain-hardening rate of the matrix. In some materials, sheets containing small voids link up the larger voids to produce final failure. Thus, the details of final coalescence, as with initiation, may vary greatly between different alloys. Given this general understanding, great improvements have been made in the (ductile) fracture toughness of specific alloys by identifying the particles responsible for void growth and eliminating their effect by changes in alloy content or manufacturing practice. However, a general theory for ductile fracture is complicated by the variety of behavior shown in the initiation and coalescence phases of fracture. As a result our subsequent discussion of ductile fracture falls primarily into two areas. One studies the growth of voids in a continuous medium subjected to prescribed stresses. The other makes use of stress and strain fields around cracks with highly simplified assumptions as to the conditions required for void growth and coalescence.

7.2   Cleavage Mechanisms

A starting point in any discussion of this topic is the classic result of Low [4], shown in Fig. 7.3. It is seen that fracture occurs only when the yield stress is reached in coarse-grained material with even higher stresses being required in fine-grained material. The implication is that plastic deformation is required to initiate a

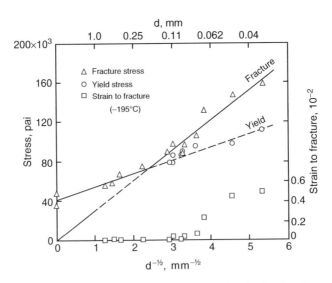

Fig. 7.3 Yield and fracture stresses at 77 K as a function of grain size for a low carbon steel. Single crystal cleavage stresses are plotted at $d^{-1/2} = 0$ (Low, ref. [4])

cleavage crack. Depending on the grain size, fracture may follow yielding or may require an increase in stress which suggests a growth-controlled rather that initiation-controlled fracture process. The first attempts to explain cleavage fracture by Zener [5], Stroh [6], and Cottrell [7] involved the pileup of dislocations at a blocked slip band to form a crack. We will not discuss these theories because careful experimental work by McMahon and Cohen [8] and McMahon [9] showed that cleavage fracture in steels is initiated by cracking of carbide particles. Steels of identical yield stress and strain-hardening rates showed cleavage with coarse carbides and ductile behavior with fine carbides. In fact, McMahon [9] concluded that "there appears to be no direct evidence of initiation of cleavage by slip band blockage in metals." Another work by Almond and Embury [10] has shown that by changing the nature of the carbides, the transition temperature may be lowered by 40 °C while increasing the yield strength by one-third. Mechanisms to explain the initiation of cleavage cracks by fracture of carbides have been proposed by Smith [11] and by Lindley, Oates, and Richards [12]. Smith's model which has received a certain amount of experimental support [13] shows that the stress for cleavage fracture is essentially independent of temperature. This result which has also been deduced from notched-bar tests, as we will discuss, supports the earlier and qualitative Ludwig-Orowan explanation of cleavage fracture discussed in Chap. 1. Smith's theory, if the dislocation contribution to the stress field is neglected, reduces to the classical Griffith equation with the crack length taken as the thickness of the grain boundary carbide and the surface energy γ replaced by a fracture energy γ_p for the ferrite matrix. Surprisingly, in view of Low's results, Smith's theory predicts no effect of grain size. This aspect is hard to pin down for the grain size and carbide size tend to change in the same way with heat treatment. The fact that the tensile stress for cleavage fracture is essentially independent of temperature greatly simplifies the explanation of the behavior of notched or cracked members. However, at very low temperatures or at high strain rates, such as near a propagating crack, fracture may be initiated by deformation due to twinning. In this case, the fracture stress may vary considerably with temperature.

7.3   Cleavage of Notched and Cracked Members

In Chap. 3 we reported the slip line solution (Figs. 3.18 and 3.19) for the plane-strain limit load of a rigid, perfectly plastic notched bar in bending. For a notch angle in Fig. 3.16 of $6.4° < \psi < 114.6°$ and a crack depth $\div$ specimen depth $a/W \geq 0.3$, the maximum tensile stress occurs below the surface and has the value

$$\sigma_y = 2\kappa \left(1 + \frac{\pi}{2} - \psi \right)$$

where κ is the yield stress in shear.

By testing specimens with varying notch angles and by determining the temperature T_{gy} at which general yield and fracture coincided, Knott [14, 15] was able to use this equation and the experimental value of κ as a function of temperature to calculate the stress σ_y for cleavage fracture. Remarkably constant values of this stress were found over a wide range of temperatures for three mild steels, and cleavage crack initiation was observed below the notch root where the tensile stress is higher than at the notch surface. Later, finite element analyses have confirmed Knott's conclusions. For an iron-37-silicon alloy from $-200\ ^{\circ}C$ to $+50\ ^{\circ}C$, little change was found in the maximum tensile stress at cleavage fracture [16].

By assuming a constant fracture stress, many aspects of slow tests on notched and cracked members may be explained and some connections can be made to impact tests. An extensive discussion of slow notch-bend tests has been given by Knott [17]. Typical behavior is sketched in Fig. 7.4. At low temperatures the uniaxial yield stress is high, only a small stress concentration is sufficient to raise the local stress to the fracture stress, and failure occurs with localized plastic deformation. At the temperature for general yield $T_{gy,}$ plastic deformation has spread over the notched cross section, but the specimen as a whole has not undergone gross deformation. Thus, T_{gy} is the limit between fractures which are

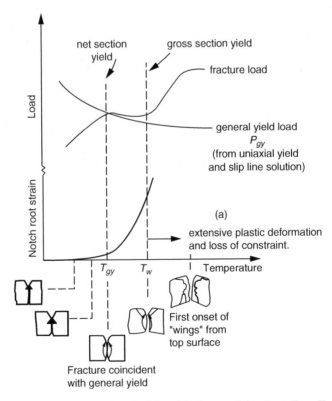

Fig. 7.4 Schematic fracture behavior of mild steel in slow notch-bend test. From Knott, ref. [17]

macroscopically "brittle" and "ductile." Above T_{gy}, the general yield, load decreases with increasing temperature due to the decrease in yield stress and strain hardening is required to elevate the local stress to the cleavage value. At T_w, yielding of the gross-section occurs, the constraint and local elevation of stress due to the notch are greatly decreased, and the fracture load rises rapidly. With increasing notch root strain, ductile void growth mechanisms may lead to fracture initiation, at or close to the notch root, at temperatures above T_{gy}. A loss of the constraining effect due to the notch is also found if the notches are not deep enough. As pointed out in Chap. 3, a ratio $a/W \geq 0.3$ is needed to ensure that constrained yield at the net section precedes yield of the gross-section. Also with thinner specimens, plane-strain conditions may not be achieved. Thus, at a given temperature, we may find a transition in fracture mode from cleavage to ductile in slow bend tests as thickness is decreased.

Dynamic testing by raising the yield stress raises the temperature for general yielding. This aspect makes it difficult to compare deeply notched slow bend specimens with the traditional Charpy V-notch impact specimen. In this case the notch depth $a/W = 0.2$ is insufficient for complete constraint. The observation that fracture may initiate at a notch by a ductile mechanism and then become cleavage may be due to increased constraint and higher stresses due to the longer crack as well as to the elevation of yield stress due to high strain rates at a growing crack.

Turning now to cleavage induced by sharp cracks rather than notches, small-scale yielding solutions are available as discussed in Chap. 3. These have been used by Ritchie, Knott, and Rice [18] to estimate K_{Ic} for a mild steel between about $-150\,^\circ$C and $-75\,^\circ$C. The problem, as they point out, is that much greater stresses can be achieved than with rounded notches. Also, the location of the maximum stress is reached close to the crack tip rather than at the elastic-plastic interface as in the slip line solutions. Since direct application of the critical tensile stress criterion for cleavage would predict failure at low nominal stress levels, these authors required that the critical stress be reached over a "characteristic" distance ahead of the tip. By trial and with plausible physical arguments, they selected a distance of two grain diameters (120 μm for their steel). Hence by knowing the fracture stress σ_f and the yield stress S_y as a function of temperature, the elastic-plastic solution for $\sigma_y(x)/S_y$ versus $x/(K/S_y)^2$ with $\sigma_y(x) = \sigma_f$ at $x = 120$ μm enables K_{Ic} to be estimated between $-150\,^\circ$C and $-75\,^\circ$C. Predictions were said to be unsuccessful at higher temperatures. Between $-75\,^\circ$C and $-45\,^\circ$C, "stable" cleavage growth preceded final failure, and at higher temperatures crack growth involved the plastic growth and coalescence of microcracks. The cleavage-fibrous transition was reported to be $+90\,^\circ$C, so there is a large temperature range from $-75\,^\circ$C to $+90\,^\circ$C remaining unexplored. It is interesting that the maximum achievable stress at 90 $^\circ$C is about 5 times the yield and approximately equal to the cleavage fracture stress measured at $-90\,^\circ$C. This suggests that above about $+90\,^\circ$C, no cleavage should be possible. Parks [19] has applied the same model to predict the effect of irradiation on a steel. He assumes the fracture stress is unchanged by irradiation with only the yield stress being altered. In this case the characteristic distance was chosen as 75 μ (approximately 3 times the prior austenite grain size).

7.4   Ductile Fracture of Notched or Cracked Members

To the present, we have discussed the variation of K_c with thickness in thin sheets using a critical strain criterion and the estimation of K_{Ic} for cleavage fracture based on a critical stress being achieved for a certain distance ahead of the crack tip. We turn now to review some of the approaches which have been taken to predict K_{Ic} or its relation to other properties when ductile fracture modes are present.

It has been noted for a given steel in different conditions of heat treatment that $K_{Ic}/n \sim$ constant where n is the strain-hardening exponent in the approximation to the stress-strain curve given by $\sigma = A\varepsilon^n$. This has suggested that local tensile instability may be involved ahead of the crack front since $\varepsilon = n$ is the strain at necking in tension. However, from elastic-plastic solutions in Chap. 3, we see that changes in n influence the stress and strain distribution and the degree of triaxial constraint near a crack tip. The strain-hardening exponent as we have seen is also of importance in void growth calculations.

One of the earliest attempts to relate fracture toughness to other properties is due to Krafft [20] who assumed that the tensile instability strain had to be reached over a distance d_T from the crack tip. He referred to this as the "process zone size." Taking a simple uniaxial model for which $\sigma_y = K_I(2\pi r)^{1/2}$ ahead of the crack and putting $\sigma_y E \varepsilon = En$; $K_{Ic} \sim En(2\pi d_T)^{1/2}$. Later estimates of d_T by observation of fracture surfaces of high strength steels have been quoted in support of this model [21]. However, since an elastic solution is combined with the condition for tensile instability based on ignoring elastic strains, the prediction can only be regarded as empirical. A problem in using this result is the difficulty of making "a priori" estimates of d_T.

Another model was proposed by Hahn and Rosenfield [22] in which crack opening displacement was used as the failure criterion. This led to

$$K_{Ic} = \left[\frac{2}{3} ES_y n^2 \varepsilon_f \right]^{1/2}$$

where ε_f is the true strain at fracture in tension. However, Jones and Brown [23] pointed out that while this prediction works quite well for AISI 4340 tempered at 600 °F, it underestimated the K_{Ic} values for higher tempering temperatures by a factor of about two.

Later work by Barsom and Pellegrino [24] has emphasized that K_{Ic} is related to the plane-strain tensile ductility and that the change in the microscopic mode of fracture from cleavage to ductile in plane-strain tension is similar to the change in crack initiation obtained in K_{Ic} tests. For steels with yield stresses from 552 to 1,720 MN/m², they propose a relation

$$K_{Ic} \sim A \sqrt{S_y \varepsilon_f^2} \quad \text{(plane strain)}$$

where A is a constant for a given steel.

Fig. 7.5 (a) Modification of slip line field in near tip region due to progressive blunting of crack tip with deformation

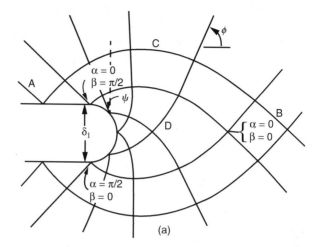

(a)

   A notable attempt to combine realistic elastic-plastic solutions for the crack tip zone with void growth models to predict ductile fracture is that of Rice and Johnson [25]. As summarized in Chap. 3, they showed that a blunted crack leads to large strains in a region extending ahead of the crack tip by a distance of approximately 2δ. It is interesting that the predicted strains within this zone, which is typically two orders of magnitude less that the plastic zone size, are almost the same for boundary conditions set for small-scale yielding or from fully plastic deep double edge notches.[2] Thus, the local strain distribution at a blunted crack may be insensitive to specimen configuration and the stresses which would be deduced from plasticity solutions for sharp cracks. Such a result would appear to be essential if δ_{Ic} or J_{Ic} are to be independent of specimen geometry even under conditions where limit loads are reached.

   Returning to the Rice and Johnson analysis, they couple this with the analysis for the growth of a spherical void due to Rice and Tracey [26] to predict the crack opening displacement at fracture δ_t in terms of the spacing of the void nuclei X_0 and their radius R_0. The results are shown in Fig. 7.5a and are based on the assumption that linkage by localized necking will occur when the distance between the blunted tip and the void is equal to the vertical radius of the now distorted, originally spherical void. Taking a cubic array of particles and observations for inclusion content of manganese sulfide in steel, they were able to show reasonable agreement between predicted and observed values of δ_T at fracture.

   Hahn and Rosenfield [27] combined the expression for crack opening displacement in plane strain from LEFM $\delta \sim K^2/2ES_y$ with the observation that crack extension will proceed when the highly strained region of distance δ ahead of the crack is equal to the space between cracked particles (between voids).

---

[2] In more recent work by Rice and his colleagues, the blunted crack model has been applied only to the case of small-scale yielding.

Since $X_0^3 = (\pi/6)(d^3/F)$ for a cubic array of particles with spacing X_0, diameter d and volume fraction (of particle) F, by setting $X_0 = \delta$.

$$K_{Ic} \sim \left[2S_yE\left(\frac{\pi}{6}\right)^{1/3}d\right]F^{-1/6}$$

Experiments have indeed confirmed the proposed dependence on volume fraction. However, fracture toughness generally decreases with increasing yield stress, and the predicted increase with particle size is contrary to intuition. However, for a constant volume fraction, the number of particles of diameter d varies as d^{-3}.

A limitation of the Hahn and Rosenfield model may be the assumption of a critical crack opening displacement δ_T equal to the interparticle spacing X_0. In recent work on aluminum alloys, Hahn and Rosenfield quote values for X_0/R_0 which are typically about 4. In this range the Rice and Johnson prediction for δ_T/X_0 is quite sensitive to the value of X_0/R_0. Experimental values for δ_T/X_0 as a function of X_0/R_0 have been given by Green and Knott [28] for a number of steels. Eight out of ten of the steels show reasonably good agreement with the Rice and Johnson prediction. In the region of low X_0/R_0 (say under 10), it could be argued that δ_T/X_0 is increasing linearly with X_0/R_0. In this case $\delta_T \sim X_0(X_0/R_0) \sim X_0F^{-1/3}$ and the particle diameter d in the Hahn and Rosenfield equation is replaced by a multiple of X_0, a more appealing result. It seems unlikely at this stage that one can represent the complicated process of ductile fracture by a single equation. We have made the modification to the Hahn and Rosenfield equation only to show that considerably different equations can result from a small change in the initial assumptions. The work of Green and Knott also points out the importance of strain hardening in the linkup of voids. They estimate values for δ at crack initiation ranging from $0.5X_0$ for non-strain-hardening materials to as much as $2.5X_0$ for strain-hardening materials. This introduces additional complexity into the question of fracture toughness prediction (Figs. 7.6, 7.7, 7.8, and 7.9).

As a final comment on ductile fracture, we should recall the reason for a size effect before turning to a completely different size effect in brittle solids. If a critical crack opening displacement is the fracture criterion, then this can be

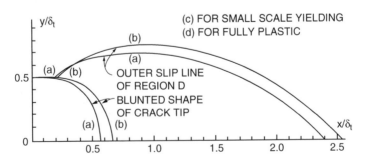

Fig. 7.6 Predicted deformed shape of crack tip and outer slip line of region D for (**a**) small-scale yielding and (**b**) fully plastic

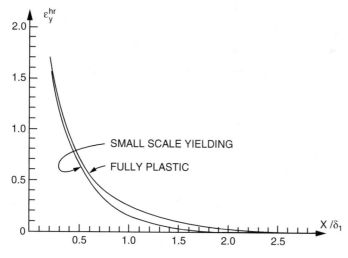

Fig. 7.7 True strain on line ahead of crack as a function of distance X of a material point from tip before deformation

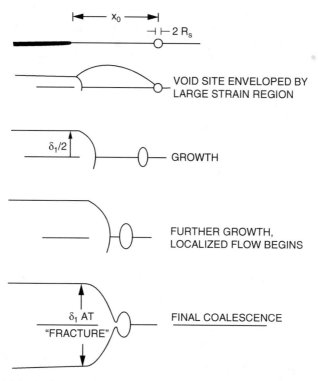

Fig. 7.8 Model for ductile fracture by the growth and coalescence of an initially spherical void with the crack tip

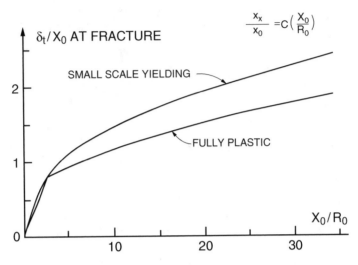

Fig. 7.9 Predicted crack opening displacement at fracture, based on the model in Fig. 7.8

provided by localized plastic deformation and small overall deformation at fracture in large specimens. However, in small specimens, gross plastic deformation may be required to produce the same displacement. Despite the differences in macroscopic behavior, the microscopic fracture mechanisms may be the same in both specimens.

References

1. Beachem CD. Microscopic fracture processes. In: Liebowitz H, editor. Fracture, vol. 1. New York and London: Academic; 1968. p. 243–349.
2. Berry JM. Cleavage step formation in brittle fracture. ASM Trans. 1959;51:556–88.
3. Burghard HC Jr, Stoloff NS. Cleavage phenomena and topographic features, "electron fractography". ASTM STP. 1967;436:32–58.
4. Low JR. The relation of microstructure to brittle fracture in "relation of properties to microstructure". Am Soc Metals. 1954;163–79.
5. Zener C. Micro mechanisms of fracture, in "fracturing of metals". Am Sol Metals. 1949;3–31.
6. Stroh AN. The formation of cracks as a result of plastic flow. Proc Roy Soc (London). 1954;223A:404–14.
7. Cottrell AH. Theory of brittle fracture in steel and similar metals. Trans AIME. 1958;212:192–203.
8. McMahon CJ, Cohen M. Initiation of cleavage in polycrystalline iron. Acta Metall. 1965;13:591–604.
9. McMahon CJ. The microstructural aspects of tensile fracture, Fundamental phenomena in the materials sciences, vol. 4. New York: Plenum; 1967. p. 247–84.
10. Almond EA, Embury JD. Instrumented impact testing of low carbon steels. Metal Sci J. 1968;2:194–200.

11. Smith E. The nucleation and growth of cleavage microcracks in mild steel. In: Proceedings of the Conference on the Physical Basis of Yield and Fracture, Inst. Phys. & Phys. Soc., Oxford; 1966. p. 36–46.
12. Lindley TC, Oates G, Richards CE. A critical appraisal of carbide cracking mechanisms in ferrite/carbide aggregates. Acta Metall. 1970;18:1127–36.
13. Oates G. Effect of temperature and train rate on cleavage fracture in a mild steel and a low-carbon manganese steel. J Iron Steel Inst. 1969;207:353–62.
14. Knott JF. some effects of hydrostatic tension on the fracture behavior of mild steel. J Iron Steel Inst. 1966;204:104–11.
15. Kontt JF. Effect of notch depth on fracture of mild steel specimens after general yield. J Iron Steel Inst. 1967;205:288–91.
16. Griffith JR, Owen DRJ. An elastic-plastic stress analysis for a notched bar in plane strain bending. J Mech Phys Solids. 1971;19:419–31.
17. Knott JF. Fundamentals of fracture mechanics. London: Butterworths; 1973.
18. Ritchie RO, Knott JF, Rice JR. On the relation between critical tensile stress and fracture toughness in mild steel. J Mech Phys Solids. 1973;21:395–410.
19. Parks DM. Interpretation of irradiation effects on the fracture toughness of a pressure vessel steel in terms of crack tip stress analysis. Trans ASME J Eng Mat Technol. 1976;98:30–6.
20. Krafft JM. Correlation of plane strain crack toughness with strain hardening characteristics of a low, a medium and a high strength steel. Appl Mat Res. 1964;3:88–101.
21. Yoder GR. Fractographic lines in maraging steels—a link to fracture toughness. Metall Trans. 1972;3:1851–9.
22. Hahn GT, Rosenfield AR. Source of fracture toughness: the relation between and the ordinary tensile properties of metals, in "applications related phenomena in titanium alloys". ASTM STP. 1968;432:5–32.
23. Jones MH, Brown WF Jr. The influence of crack length and thickness in plane strain fracture toughness tests, in "review of developments in plane strain fracture toughness testing". ASTM STP. 1970;463:63–101.
24. Barsom JM, Pellegrino JV. Relationship between and plane strain tensile ductility and microscopic mode of fracture. Eng Fract Mech. 1973;5:209–21.
25. Rice JR, Johnson MA. The role of large crack tip geometry changes in plane strain fracture. In: Kanninen MF, et al., editors. Inelastic behavior of solids. McGraw-Hill Book Co., Inc.; 1970. p. 641–72.
26. Rice JR, Tracey DM. On the ductile enlargement of voids in triaxial stress fields. J Mech Phys Solids. 1969;17:201.
27. Hahn GT, Rosenfield AR. Metallurgical factors affecting fracture toughness of aluminum alloys. Metall Trans. 1975;6A:653–70.
28. Green G, Knott JR. The initiation and propagation of fracture in low strength steels. Trans ASME J Eng Mat Technol. 1976;98H:37–46.

Chapter 8
The Fracture of Brittle Solids

8.1 Introduction

Traditionally, mechanical engineering design has been based almost entirely on the use of ductile metals. They are tough, are relatively strong, and can readily be formed into complicated shapes. A great deal of research has been devoted to the processing and mechanical behavior of these materials, and design methods for their use are rather well established. By contrast, relatively little attention has been given, in the past, by mechanical engineers, to materials showing brittle behavior. Civil engineers have used such materials, but generally in situations where they are in compression and with generous safety factors being allowed. In recent years, however, there has been an increasing interest in the use of brittle solids in engineering design. Perhaps the principal reason for this is that the materials which have the greatest strength at very high temperatures are brittle at ambient temperature, for example, graphite, aluminum oxide, beryllium oxide, silicon carbide, and silicon nitride. If such materials could be fabricated and used, reliably, the economic advantages would be considerable in many high-temperature applications. The usefulness of brittle materials is, of course, not confined to elevated temperatures. In many other situations, the properties of specific brittle materials (e.g., wear and corrosion resistance, high compressive strength, low neutron absorption in the case of BeO, transparency in the case of quartz and glass, etc.) make them attractive to the designer.

As we will discuss, the mechanisms of tensile and compressive failure in brittle solids is quite different. In tests on bulk specimens, the compression strength is usually at least an order of magnitude greater than the tensile strength. For this reason, a major objective of the designer is to keep brittle materials in compression wherever possible. However, in many situations tensile stresses are unavoidable and provide a limiting factor in design.

Another reason for interest in brittle solids is the expression for the theoretical strength, for the materials with the largest values of Young's modulus are brittle at

© Springer Science+Business Media New York 2016
C.K.H. Dharan et al., *Finnie's Notes on Fracture Mechanics*,
DOI 10.1007/978-1-4939-2477-6_8

ambient temperature. For example, the modulus for aluminum oxide is more than twice that of steel and the modulus to density ratio is more than five times greater. Graphite fibers have been prepared with a modulus over three times that of steel and a modulus to density ratio over 12 times that of steel.

It is amusing to speculate on the changes which might arise in engineering design if these high theoretical strengths could be attained. For example, an alumina flywheel used at the theoretical strength could store as much energy in a given volume as gasoline. The thought of using this approach to solve the automobile smog problem in a city such as Los Angeles is tempered by the realization that failures at such high stress levels might also solve the population growth problem. A factor often ignored when high theoretical strength levels are discussed is that a strength of $E/10$ requires a strain of about 10 %, and such large strains would require major changes in engineering design. For example, Gordon [1] has pointed out that in present jet transports, a strain of about 1.5 % in the main wing spans would mean that the wing tips were vertical, while at 3 % strain the wing tips would touch over the pilot's head! Obviously, this is an extreme example and not something that one has to worry about in working with bulk specimens of brittle solids for, typically, they fail in tension or bending at strains of about 0.001, i.e., at a stress level $\sigma \cong E/1,000$. However, microscopic whiskers (with cross section of about 10^{-4} in. $\times 10^{-4}$ in.) have been prepared with strength levels approaching the theoretical value and these may be bent elastically like the hypothetical aeroplane wing. Eventually, the use of these whiskers in structural materials may mean that large strains will have to be considered by the designer.

Putting aside these optimistic speculations related to the theoretical strength and turning back to present day reality, we have to accept the fact that structural parts made of brittle solids fail in tension or bending at about 1 % of their theoretical strength. Usually there is considerable variability in strength with nominally identical specimens failing at widely different stress levels. Also, there is a "size effect" on strength, for as the specimen size is decreased, the average stress at fracture tends to increase. These observations may be explained by noting that all real materials will inevitably contain a distribution of flaws of varying severity such as surface scratches or steps, cracks at grain boundaries, and inclusions. If the material behavior is predominantly elastic, the high local stresses around such flaws cannot be relieved appreciably by plastic deformation as they could be if ductile behavior occurred. Thus, it is not unreasonable to find that strength values for such materials as glass, graphite, and aluminum oxide are only a small fraction of their theoretical value. The variability in strength[1] is consistent with the idea that some specimens will contain more severe flaws than others while the "size effect" follows from the decreased probability of finding a severe flaw in a small specimen. We will

---

[1] Variability in strength is an inherent feature of brittle solids although it may also be influenced by faulty testing techniques. One should be cautious in evaluating novel techniques which purport to reduce variability in testing brittle solids. Such tests have been reported in the literature based on a very small number of specimen and with high and low strength values having been discarded by an overly helpful laboratory assistant.

examine the quantitative implications of these observations for failure under tensile states of stress in some detail before turning to consider failure in compression.

8.2 Failure Under Tensile States of Stress

By assuming a flaw of a given geometry, for example, a thumbnail crack, and knowing fracture toughness values, estimates may be made for the flaw sizes which are responsible for the observed tensile strength values of brittle solids. When this is done it is found that typical flaws in such materials as glass, alumina, or magnesia are too small to be detectable by present experimental techniques.[2]

Thus, the deterministic linear elastic fracture mechanics approach by which one relates strength to known flaw size cannot be applied at present to predict the strength of these brittle solids. Rather, one has to take a probabilistic approach and obtain a distribution of strength values from, say, 30 to 50 tests on nominally identical specimens under a given type of loading. Then, following the procedures first suggested by Weibull, the probability of failure may be predicted for parts of various sizes under other loading conditions.

Before discussing the approach developed by Weibull [2], it may be useful to give some background on methods of treating data that show scatter. This information is useful not only in studying fracture but also in treating fatigue, reliability, and other problems.

The behavior of a brittle solid in tension tests may be described by plotting the histogram shown in Fig. 8.1. By testing enough specimens and taking smaller

---

[2] By contrast to polycrystalline ceramics where the strength impairing flaws are distributed throughout the volume of a part, in glass the flaws are apparently confined to the surface. Very few cases of internal failure in glass have been reported. Carefully prepared glass rods are greatly degraded in strength when subjected to surface damage such as sandblasting. On the other hand, etching away a surface layer with hydrofluoric acid, flame polishing, or the presence of residual compressive stresses may greatly increase the strength of glass parts. These observations, the variability in strength, and the discrepancy between observed strengths and the theoretical value are strong indirect evidence for the presence of "Griffith" cracks. However, to our knowledge there has, as yet, been no direct observation of the inherent flaws in glass.

Taking $G_{Ic} \cong 0.03$ in $-$ lb/in^2, $E = 10^7$ psi, and the strength of such glass as 10^4 psi, from the thumbnail flaw solution, we would expect a maximum crack depth of about 10^{-3} in if the cracks were shallow scratches. The resolution R of an optical microscope is given by $R \cong$ Wavelength of light/(refracture index, $\sin U$) where $2U$ is the solid angle at which light passes into the objective lens. In the best case with $\sin U \cong 0.95$, violet light and oil immersion, $R \cong \frac{6 \times 4 \times 10^{-5}}{1.6 \times 0.95} \cong 1.6 \times 10^{-5}$ cm $= 0.16\mu$.

Thus we conclude that the crack faces must be closed. Presumably this accounts for the lack of success of electron microscope observations since the replicas could then not penetrate the cracks.

However, what has been done is to expose the glass at elevated temperature to sodium vapor. A thin layer of high expansion coefficient glass is formed, and on cooling, tensile stresses are produced which form a crack pattern like drying mud. This pattern appears to initiate at preexisting surface flaws for if the glass is etched first with hydrofluoric acid, no crack pattern appears.

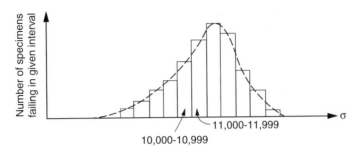

Fig. 8.1 Histogram

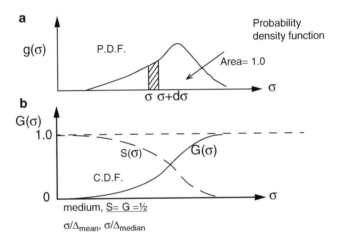

Fig. 8.2 (**a**) P.D.F probability density function, (**b**) C.D.F cumulative density function

intervals, we can draw, eventually, the continuous curve shown dashed in Fig. 8.1. This curve then provides an approximation to the distribution of strength of all specimens of the given material in the particular conditions of grain size, surface, temperature, etc., for which the tests were carried out. By scaling the ordinate to make the area under the curve unity, we obtain the probability density function $g(\sigma)$ shown in Fig. 8.2a. This shows the relative frequency with which different values of strength are observed. The shaded area in Fig. 8.2a represents the probability that strength observations fall in the interval σ to $\sigma + d\sigma$. Clearly, the entire area under the curve is taken as unity since this is the probability that a specimen will have a strength between zero and infinity. Now we ask for the probability $G(\sigma)$ that a specimen will have failed at a stress level $\leq \sigma$. This probability increases monotonically from 0 to 1 as σ increases and is merely the area under the $g(\sigma)$ curve from zero to σ.

Thus $G(\sigma)$ the cumulative distribution function shown schematically in Fig. 8.2b is given by

$$G(\sigma) = \int_0^\sigma \sigma g(\sigma) d\sigma$$

The probability that a specimen survives a stress level is $S(\sigma) = 1 - G(\sigma)$. Several properties of the strength distribution are often of interest. The mean for the histogram is

$$\sigma_m = \frac{\sum \sigma_i N_i}{\sum N_i}$$

where σ_i = mid point of ith interval and N_i = number of specimens failing in that interval. For a continuous distribution it is

$$\sigma_m = \int_0^\infty \sigma g(\sigma) d\sigma$$

Since

$$g(\sigma) = \frac{dG(\sigma)}{d\sigma} = -\frac{dS(\sigma)}{d\sigma}$$

$$\sigma_m = \int_0^1 \sigma dG = -\int_1^0 \sigma dS = -\{(\sigma \to \infty)\,(S \to 0) - (0)\,(1)\} + \int_0^\infty S d\sigma$$

$$\left(= -\left\{\sigma S(\sigma)_{S=1}^{S=0}\right\} + \int_0^\infty S d\sigma \right)$$

$$= \int_0^\infty S d\sigma = 0 \text{ since } S \to 0 \text{ more rapidly than } \sigma \to \infty$$

The mode (most likely value)

$$\sigma_{\text{mode}} = \frac{dG(\sigma)}{d\sigma} = \frac{d^2G(\sigma)}{d\sigma^2} = 0$$

The median (middle value)

$$\int_0^{\sigma_{\text{median}}} g(\sigma) d\sigma = \frac{1}{2} = \int_{\sigma_{\text{median}}}^\infty g(\sigma) d\sigma$$

The variance a^2 (the square of "a" the standard deviation) is

$$a^2 = \sum_{i=1}^{N} \frac{(\sigma_i - \sigma_m)^2}{N-1} \qquad \text{for } N \text{ observation or}$$

$$a^2 = \frac{\sum_{i=1}^{N} N_i (\sigma_i - \sigma_m)^2}{(N-1)} \qquad \text{for the histogram}$$

where σ_i is the midpoint of an interval and N_i the number of observations in the interval. The variance for the continuous distribution is

$$a^2 = \int_0^{\infty} (\sigma - \sigma_m)^2 g(\sigma) d\sigma$$

With enough observations one could obtain $g(\sigma)$ by plotting a histogram as shown in Fig. 8.1. However, usually we have a limited number of observations, and it may be more convenient to go directly to the cumulative distribution function $G(\sigma)$. If a group of N observations has been obtained, we first rank these in order of increasing strength $j = 1, 2, \ldots, N$ where $\sigma_j \geq \sigma_{j-1}$. In attempting to plot a curve of $G(\sigma)$ versus σ, there is no problem in plotting the abscissa; it is merely the observed strength σ_j for each specimen. However, the value of the ordinate $G(\sigma)$ to be assigned to each specimen is less clear. Intuitively, we would expect to plot the point for the weakest specimen $j = 1$ in the interval 0 to $1/n$, the next in the interval $1/N$ to $2/N$, and in general $(j-1)/N < G(\sigma) < j/N$. When N is large there is no problem for $G(\sigma) \cong j/N$, but for small N, we cannot choose $G(\sigma) = (j-1)/N$ or $G(\sigma) = j/N$ for this would mean that we have found either the weakest or strongest of all possible specimens in our limited sampling. One approach is to represent each data point by a bar from $(j-1)/N$ to j/N and draw the curve of $G(\sigma)$ versus σ through these bars. A more detailed consideration of the plotting position corresponding to the jth observation becomes rather involved. Expressions often used are

$$G(\sigma) = (j-1)/N \qquad \text{and} \qquad G(\sigma) = \frac{j - 0.3}{N + 0.4}$$

The first is the mean value of probability corresponding to the jth observation. The second expression is a close approximation to the median plotting position. That is, half of the jth observations are above and half are below this value.

It is often convenient for subsequent calculations to fit an analytical expression to the distribution $g(\sigma)$ or $G(\sigma)$. A number of different distributions are described in texts on probability. In selecting between these it is useful to have some physical understanding of the phenomenon being measured and of the situations in which the different distributions might be expected to arise.

In testing materials we can visualize two idealized models, the parallel and the series. In the first the strength depends on the combined parallel effect of many individual units, e.g., the strands of a wire cable. In the second the strength is that of the weakest unit, e.g., as in a chain. It is known that when a result depends linearly on a large number of noninteracting variables, all more or less equally important in their effect, then the resulting distribution function is the normal distribution. For example in studying the yielding of a polycrystalline metal where all grains in a given cross section participate in yielding, we might expect this distribution. To be more precise, for a parallel model, it may be shown that provided $S(\sigma)$ for the individual elements approaches zero more rapidly than $1/\sigma$, then the resulting strength distribution of the assembly is normal [3].

Turning to the series (or chain) model, this would seem to provide a more realistic description of the failure of brittle solids under tensile states of stress and of fatigue. We can see the type of distribution that would be expected by considering a chain of N links in which the individual failure probabilities (at a given load) are $F_1, F_2, \ldots, F_i, \ldots, F_N$. Letting G and S denote the failure and survival probabilities of the chain as a whole, we can write the probability that the chain survives as the product of the individual survival probabilities. Thus (with π denoting a product and Σ a sum over N terms),

$$S = 1 - G = (1 - F_1)(1 - F_2) \ldots (1 - F_i) \ldots (1 - F_N)$$

$$1 - G = \prod_{i=1}^{N}(1 - F_i)$$

$$\ln(1 - G) = \sum 1 - F_i \cong -\sum F_i \quad \text{(if small)}$$

$$G = 1 - e^{-\Sigma F_i}$$

Now, we wish to generalize this result to a volume V composed of elements dV which may be under different tensile stresses σ. In the case of glass we let V represent the area under tension rather than the volume. We neglect the effect of compressive stress and consider only uniaxial tensile stresses, although the approach can be extended to multiaxial stress. It is reasonable to assume that F_i for each volume element dV is proportional to its volume and some function of stress. Writing $F_i = \phi(\sigma)dV$, the preceding equation becomes

$$G(\sigma) = 1 - e^{-\int \phi(\sigma)dV}$$

The argument we present was first put forward by the Swedish engineer Weibull [2] in 1938 and he chose $\phi(\sigma) = (\sigma/\sigma_0)^m$ as a simple empirical expression which fitted a large amount of experimental data. Thus,

$$G(\sigma) = 1 - e^{-\int \phi(\sigma/\sigma_0)^m dV} = 1 - e^{-B}$$

where B, which was termed "the risk of rupture" by Weibull, is merely an abbreviation for $\int (\sigma/\sigma_0)^m dV$. The parameter σ_0, which has the dimensions of stress is a scaling constant, while the parameter m is often referred to as the Weibull parameter or the flaw density parameter. We note that as $m \to \infty$, $G = 0$, $\sigma < \sigma_0$; $G = 1$, $\sigma > \sigma_0$ so there is no variability in strength. As m decreases, the variability in strength increases. Values which have been quoted for m and which can only be taken as typical are:

Glass	$m = 8$
MgO	$m = 12$
Al_2O_3	$m = 15$
Tool steel quenched to martensite	$m = 20$
Steel in liquid air	$m = 24$

For the case of volume V subjected to uniform uniaxial tension

$$G(\sigma) = 1 - e^{-V(\sigma/\sigma_0)/m}$$

If we consider specimens of two sizes V_1, V_2, with stresses σ_1, σ_2, then

$$G_1(\sigma) = 1 - e^{-V(\sigma_1/\sigma_0)^m}, \quad G_2(\sigma) = 1 - e^{-V(\sigma_2/\sigma_0)^m}$$

For an equal probability of failure $G_1 = G_2$ and so $(\sigma_1/\sigma_2) = (V_2/V_1)^{1/m}$. This allows us to scale test data for size as long as we know the exponent m.

A limitation of the simple two-parameter Weibull distribution is that it predicts a small but finite probability of failure at low stresses. In brittle materials containing residual stresses, this may not be unrealistic, but in other cases this result violates our physical intuition. To take care of this objection, Weibull proposed a three-parameter distribution:

$$G_1(\sigma) = 1 - e^{-\int \left(\frac{\sigma - \sigma_u}{\sigma_0}\right)^m dV} \quad \sigma > \sigma_u$$
$$= 0 \quad \sigma \le \sigma_u$$

The additional parameter σ_u is often referred to as the zero strength. It is important to note that in describing a given set of data with both the two-parameter and the three-parameter distributions, the two values of m and σ_0 will be quite different. A result which can be obtained by comparing the coefficients of variation (standard deviation ÷ mean) for the two and three-parameter distributions is $m_{\sigma_u} = m_{(\sigma_u=0)}(1 - \sigma_u/\sigma_{mean})$. The three-parameter distribution may in principle provide more accurate prediction than the two-parameter distribution. However, the parameters themselves are not as easily obtained and analytical calculations are usually no longer possible. For this reason we shall use the two-parameter distribution in much of our future discussion.

8.3   Estimation of Parameters

The estimation of the parameters of a distribution and the choice between different distributions are discussed at length in the literature. Here we take the simple approach suggested by Weibull of plotting the distribution in a form which yields a linear plot.

For a volume V in uniaxial tension at stress level, or an area in the case of glass, the two-parameter distribution is

$$G(\sigma) = 1 - e^{-V(\sigma/\sigma_0)^m V}$$

or

$$\ln \frac{1}{1 - G} = \left(\frac{\sigma}{\sigma_0}\right)^m V$$

Writing $G = j/(N + 1)$, the mean rank and taking logarithms again

$$\log\ln\left(\frac{N + 1}{N + 1 - j}\right) = m\log\sigma - m\log\sigma_0 + \log V$$

or

$$\log\,\log\left(\frac{N + 1}{N + 1 - j}\right) = m\log\sigma - m\log\sigma_0 + \log V + \log\,\log e$$

Thus a plot of $\log[(N + 1)/(N + 1 - j)]$ versus σ on log-log paper should establish the value of m. For the three-parameter distribution, the term $m \log \sigma$ is replaced by $m\log(\sigma - \sigma_u)$. Since σ_u is not known, we choose trial values and select the one for which a plot of $\log[(N + 1)/(N + 1 - j)]$ versus $(\sigma - \sigma_u)$ on log-log paper best fits a straight line.

Although the tension test is the easiest to analyze, it is one of the most difficult to carry out with precision. Turning to the pure bending test of a specimen containing a volume distribution of flaws, we consider the bending specimen of Fig. 8.3 and integrate from the fiber at which $\sigma = \sigma_u, y = h(\sigma_u/\sigma_b)$ to the outer fiber $y = h$. Then B the risk of rupture becomes

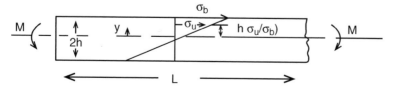

Fig. 8.3 Pure bending on specimen of volume V_b

$$B = \frac{V_b}{2(m+1)}\left(1 - \frac{\sigma_u}{\sigma_b}\right)\left(\frac{\sigma_b - \sigma_u}{\sigma_0}\right)^m$$

and we can treat the expression $G(\sigma_b) = 1 - e^{-B}$ in the same way as tension data to obtain

$$\log\,\log\left(\frac{N+1}{N+1-j}\right) = (m+1)\log(\sigma_b - \sigma_u) - \log\sigma_b + \log\frac{V_b}{2(m+1)} + \log\,\log e$$

So we plot $\log\,\log[(N+1)/(N+1-j)] + \log\sigma_b$ against $\log(\sigma_b - \sigma_u)$. The value of σ_u is again obtained by trial until a straight line of slope $(m+1)$ is obtained.

Several other simple methods for estimating the parameters are available for a two-parameter Weibull distribution. For example, it may be shown that the "coefficient of variation"

$$\frac{a}{\sigma_m} = \frac{\text{standard deviation}}{\text{mean}} = \left\{\frac{\Gamma\left(1+\frac{2}{m}\right)}{\Gamma^2\left(1+\frac{1}{m}\right)} - 1\right\}^{1/2} \cong \left(\frac{1}{m}\right)^{0.94} \text{ to an excellent approximation}$$

$\left(\frac{1.2}{m}\right)$ to a fair approximation.

An estimate of m may also be made form the extreme values of the distribution. Conversely, knowing m it is often useful to estimate extreme values.

Assuming that $G(\sigma)$, the probability that a stress level $\leq \sigma$ causes failure in a specimen, is known, we ask for the probability that N specimens all survive at a stress level σ. This probability is $[S(\sigma)]^N = [1 - G(\sigma)]^N$. Thus, the probability $H(\sigma)$ that a series of N specimens will contain one or more failures when loaded to stress level σ is

$$H(\sigma) = 1 - [S(\sigma)]^N = 1 - 1[1 - G(\sigma)]^N$$

The corresponding probability density function is given by

$$h(\sigma) = dH\sigma/d\sigma.$$

The most likely value of this distribution of the weakest values is given by $dh/d\sigma = 0$. For the two-parameter case, the result is

$$\frac{\sigma_{\min}}{\sigma_{\text{mean}}} = \left(\frac{m-1}{mN}\right)^{1/m}\frac{1}{\Gamma(1+1/m)}$$

This value, the mode of the extreme value distribution, is shown schematically in Fig. 8.4. We note that $h(\sigma)$ depends on the number of specimens being considered. If we know $G(\sigma)$ [or G] and N, then the most likely value of the strength of the weakest specimen may be estimated.

On the other hand, we may wish to estimate the most likely value for the largest of N observations (earthquakes, floods, flaw sizes). In this case, the probability that all

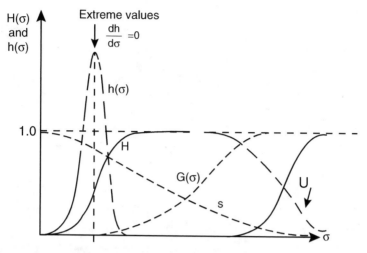

Fig. 8.4 Distribution of extreme values in batch of size n

specimens fail at a stress level $\geq \sigma$ is $[G(\sigma)]^N$. Thus, the probability U that the strength of N specimens contain at least one (one or more) survivor at stress level σ is

$$U(\sigma) = 1 - [G(\sigma)]^N$$

Taking the second derivative of $U(\sigma)$ and setting it equal to zero gives the most likely strongest value in a batch of size N. For the two-parameter Weibull distribution, extreme value predictions can be given in the concise form [4] shown in Fig. 8.5 and can be used to estimate m.

As long as we work with the two-parameter distribution, we can apply the coefficient of variation or extreme value method of estimating m to any type of test. In fact we can pool data for different sizes of specimens and different types of tests provided we normalize each strength observation by the mean (or the median) for that size of specimen and type of test.

To show this, we write the stress σ in terms of a reference stress σ_r (say the maximum stress) and location coordinates, i.e., $\sigma = \sigma_r f(x, y, z)$. So, the survival probability may be written as

$$S(\sigma_r) = e^{-(\sigma_r/\sigma_0)^m \int f^m(x, y, z)\, dxdydz}$$

or

$$\frac{\ln \frac{1}{S}}{(\sigma_r)^m} = \left[\frac{1}{\sigma_0^m} \int f^m(x, y, z) dxdydz \right]$$

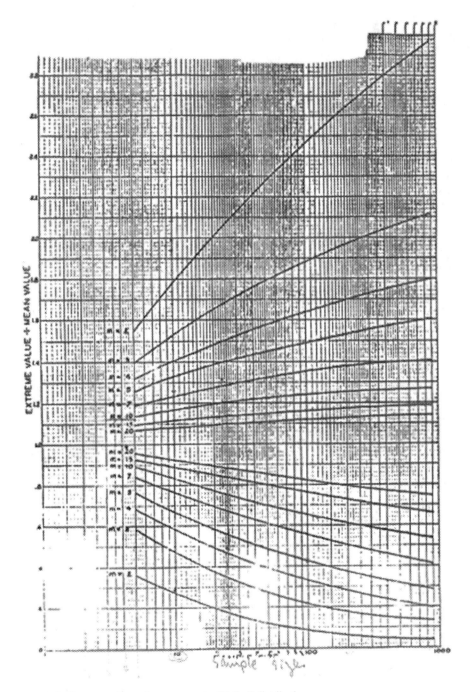

Fig. 8.5 Extreme values of two-parameter Weibull distribution

Thus for a given survival probability S' and corresponding reference stress σ_r' the value of the quantity in square brackets is known. The two-parameter distribution may then be written as

$$S = e^{-\left(\ln 1/S'\right)\left(\sigma_r/\sigma_r'\right)^m}$$

Hence, we obtain the Weibull distribution for a normalized stress (σ_r/σ_r') by choosing the appropriate value of S'. If we normalize all stresses by the median, then $S' = 1/2$; $\ln 1/S' = 0.693$ and $S = e^{-0.693\left(\sigma_r/\sigma_r'\right)^m}$ where $\sigma_r' = $ median strength for a given size specimen in a given type of test. Similarly, if we normalize by the mean strength,

$$S = e^{-\Gamma^m(1+1/m)\left(\sigma_r/\sigma_r'\right)^m}$$

Detailed treatments of the estimation of parameters are available in the literature. We have presented here only a few simple procedures that are often adequate in practical problems. After a particular distribution has been selected, its "goodness of fit" can be evaluated by the standard chi-squared test on confidence limits placed on the parameters [5].

It is also possible to get a feeling for the variability involved in estimating m from small samples by carrying out experiments with the computer. By selecting a Weibull distribution with given parameters and then calling for a series of N random numbers from 0 to 1 (i.e., probabilities), a set of N observations corresponding to a known Weibull distribution is obtained. The parameters can be estimated (as if they were unknown) and the estimates compared with the known value. Doing this a number of times, the mean m and the standard deviation of m are obtained. It is found that (St. Dev. of $m \div$ mean m) $\cong 1/N^{1/2}$ with no single method of estimating the parameter m being greatly superior to the others [4].

8.4   Some Applications of the Weibull Distribution

One of the attractions of Weibull's procedure is that it allows us to compare fracture tests of different sizes and stress distributions. This is useful, for example, in predicting the behavior of large parts from smaller laboratory size specimens. We already saw that for two tension specimens volumes V_1 and V_2, the corresponding stresses for equal probability of failure, using the two-parameter distribution, were related by

$$\frac{\sigma_1}{\sigma_2} = \left(\frac{V_r}{V_1}\right)^{1/m}$$

From what we have said about pooling of data by scaling the strength values, we see that this result applied to any geometrically similar specimens where the stress state can be described by a reference stress multiplied by a function of the coordinates.

To compare bending tests and tension tests, we recall that the "risk of rupture" in discussing estimation of parameters was

$$B = \frac{V_b}{2(m+1)} \left(1 - \frac{\sigma_u}{\sigma_b}\right) \left(\frac{\sigma_b - \sigma_u}{\sigma_0}\right)^m$$

for pure bending, while for tension it is merely

$$B = V \left(\frac{\sigma - \sigma_u}{\sigma_0}\right)^m.$$

Equating the two expressions for B, we obtain the relation between σ and σ_b for the same probability of failure:

$$\frac{\sigma}{\sigma_b} = \left[\frac{1}{2(m+1)} \frac{V_b}{V_t}\right]^{1/m} \left(1 - \frac{\sigma_u}{\sigma_b}\right)^{1+1/m} + \frac{\sigma}{\sigma_b}$$

This type of prediction explains why higher mean strength values are obtained in bending than in tension. Physically of course this arises because a very small volume of material is exposed to the highest stress in a bending test (Fig. 8.6). The corresponding probability of encountering a severe flaw at the location of the highest stress is thus very small. As a result in stress states which involve extreme stress gradients (e.g., thermal shock tests [6]), the Weibull approach will predict higher strength values than classical theory.

Another case in which Weibull's approach leads to useful predictions is the case of a spherical punch pressed against a large glass plate as shown in Fig. 8.7. At a certain load, a crack forms as a ring on the surface and flares out below the surface to form the frustum of a cone. Increasing the load causes the crack to grow. With the spherical indenter the contact area grows also, and additional ring cracks may form (Fig. 8.7).

The stresses in the glass prior to fracture are known from elastic theory, and since glass inevitably fails from surface flaws, the important aspect of the stress state is the tensile stress on the surface. Under the indenter all principal stresses are compressive, while on the free surface the circumferential stress is compressive and the radial stress is tensile with the values shown in Fig. 8.7. The radial stress outside the contact area is given by

$$\sigma_r = \frac{1 - 2\nu_{\text{glass}}}{2\pi} \frac{P}{a^2} \left(\frac{a}{r}\right)^2 \quad r \geq a.$$

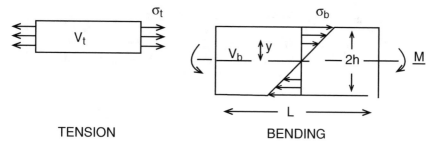

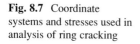

TENSION   BENDING

Fig. 8.6 Comparison of tension and bending on brittle solids

Fig. 8.7 Coordinate
systems and stresses used in
analysis of ring cracking

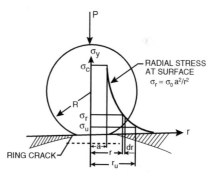

Thus, if fracture occurred when the radial tensile stress reaches a critical value, we would expect the load P for fracture to be related to indenter radius by $P \sim R^2$. Also, we would expect no scatter in the load at fracture with all fractures starting at the rim of the contact area. However, as with all tests on brittle solids, the fracture load and the fracture location show considerable scatter. There is an effect of size on strength with the mean value of the stress σ_a at the rim of the contact area, increasing as the indenter size is decreased. A prediction of these effects using Weibull's approach has been given in the literature [7].

Safety Factor

We now ask for given values of the parameters, what do we mean by safety factor when we come to design with a brittle material? As safety factor we take mean strength/design stress. For simplicity, we consider uniaxial tension and the mean is given by

$$\sigma_m = \sigma_u + \sigma_0 V^{-1}\Gamma(1 + 1/m)$$

where the gamma function which is defined as $\Gamma(n) = \int_0^\infty x^{n-1}e^{-x}dx$ may be looked up in mathematical tables. (For $1 \le n \le 2$, $0.22 \le \Gamma \le 1$.)

The design stress must be written in terms of a probability of failure

$$G(\sigma) = 1 - e^{-\left(\frac{\sigma - \sigma_u}{\sigma_0}\right)^m V}$$

$$-\ln(1 - G) = \left(\frac{\sigma - \sigma_u}{\sigma_0}\right)^m V$$

or

$$\sigma = \sigma_u + \sigma_0[-\ln(1 - G)]^{1/m} V^{-1/m}$$

Thus

$$\text{Safety Factor} = \frac{\sigma_m}{\sigma} = \frac{\sigma_u + \sigma_0 V^{-1/m}\Gamma(1 + 1/m)}{\sigma_u + \sigma_0 V^{-1/m}[-\ln(1 - G)]^{1/m}}$$

$$\text{for } \sigma_u = 0: \quad \frac{\sigma_m}{\sigma} = \Gamma(1 + 1/m)/[-\ln(1 - G)]^{1/m}$$

and we may plot σ_m/σ as a function of m for various values of failure probability G or survival probability $S = 1 - G$. This is shown in Fig. 8.8. If m is low, very large safety factors are needed to ensure a high survival probability. Of course, when we start talking about very high survival probabilities, such as 99.999 %, the discussion becomes somewhat academic. Predictions are only realistic if we know the underlying strength distribution to a high level of confidence.

One solution to the problem of high safety factors is to proof test and eliminate the weaker specimens. On the assumption that proof testing does not degrade the strength of the survivors, we can describe the truncated distribution $G_p(\sigma)$ shown in Fig. 8.9 by

$$G_p(\sigma) = \frac{\text{Number of failures (of survivors of proof test) at stresses} \le \sigma}{\text{Number surviving proof test}}$$

$$G_p(\sigma) = \frac{NG(\sigma) - NG(\sigma_p)}{N - NG(\sigma_p)} = \frac{G(\sigma) - G(\sigma_p)}{1 - (\sigma_p)} \qquad \sigma \ge \sigma_p$$

$$= 0 \qquad\qquad\qquad\qquad\qquad \sigma < \sigma_p$$

If damage is induced by proof testing we have, essentially, a case of low cycle fatigue. The problem is hard to treat quantitatively but data may be presented as shown in Fig. 8.10.

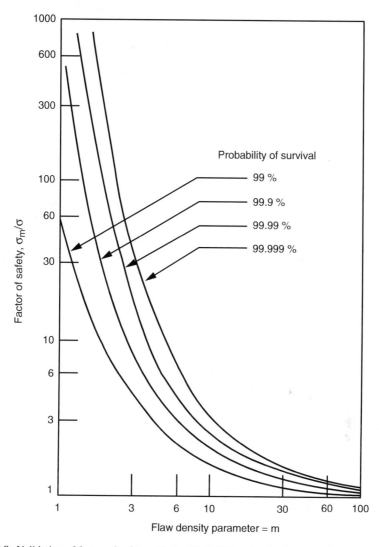

Fig. 8.8 Validation of factor of safety with the Weibull flaw density fracture with $\sigma_u = 0$

8.5 The Location of Fracture in Brittle Solids

This is an aspect which has only recently received attention in fracture studies. In the ring cracking of glass plates under spherical indenters and in the three-point bending of beams, for example, the location of fracture is easily observed and shows considerable scatter. In the case of the ring cracks the average fracture location is well removed from the location of the maximum tensile stress. A direct

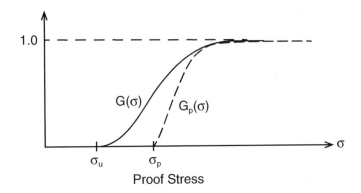

Fig. 8.9 Original and truncated distributions

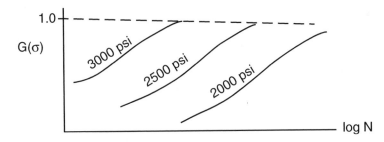

Fig. 8.10 Cumulative distribution of fracture probability as a function of the cycles N to a given stress level

application of Weibull's approach gives no information about fracture location because it provides only the failure probability of the specimen as a whole. What we have to do is to consider small differential regions in the specimen and determine the probability that one region fails while the others survive. Computations have been carried out for single and multiple ring cracks, beam bending, and wave propagation in slender rods [7, 8].

To illustrate one possible use of fracture location observations, we consider the propagation of a rectangular compressive pulse down a slender glass rod. As the pulse meets the free end of the rod shown in Fig. 8.11, it reflects as tension. Thus, starting at a distance $L/2$ from the free end, a tensile pulse sweeps down the length of the rod and also sweeps down the segment of length $L/2$. For convenience, we ignore cases in which fracture occurs in the $[0, L/2]$ segment and measure distance down the rod as shown in Fig. 8.11. We write the probability of failure occurring in length x as

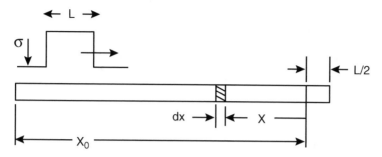

Fig. 8.11 Coordinates used to describe fracture location in a slender rod due to rectangular stress pulse

$$G(\sigma) = 1 - e^{-\left(\frac{\sigma-\sigma_u}{\sigma_0}\right)^m Ax}$$

where A is the surface area associated with unit length of rod. Hence, the probability that length x survives and that failure occurs in dx is

$$e^{-\left(\frac{\sigma-\sigma_u}{\sigma_0}\right)^m Ax}\left[1 - e^{-\left(\frac{\sigma-\sigma_u}{\sigma_0}\right)^m Adx}\right]$$

Since dx is small, we can replace the exponential in the second bracket by the first two terms of the series and get the probability that length x survives and length dx fails as

$$e^{-\left(\frac{\sigma-\sigma_u}{\sigma_0}\right)^m Ax}\left(\frac{\sigma-\sigma_u}{\sigma_0}\right)^m Adx$$

To get a cumulative distribution function, we integrate

$$P(x) = -e^{-\left(\frac{\sigma-\sigma_u}{\sigma_0}\right)^m Ax}\Big]_0^x = 1 - e^{-\left(\frac{\sigma-\sigma_u}{\sigma_0}\right)^m Ax} = 1 - e^{-\lambda x}$$

where $\lambda = [(\sigma - \sigma_u)/\sigma_0]^m A$ is a constant if σ is held constant. If the rods are very long, then as $x \to \infty$, $P(x) \to 1.0$. For a rod of length x_0, measured as shown in Fig. 8.11, we can write the conditional distribution of fracture length, given that fracture has occurred in the interval $[0, x_0]$ as

$$\frac{1 - e^{-\left(\frac{\sigma-\sigma_u}{\sigma_0}\right)^m Ax}}{1 - e^{-\left(\frac{\sigma-\sigma_u}{\sigma_0}\right)^m Ax_0}}$$

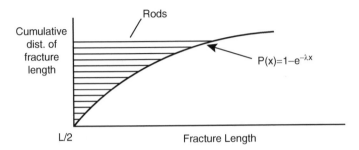

Fig. 8.12 Use of fracture length specimen to illustrate a probability distribution

In this way we deal with probabilities from zero to unity. Experiments show good agreement with the predicted exponential distribution of fracture length. The specimens themselves can be ranked in order of increasing fracture length to portray the probability distribution as sketched in Fig. 8.12. By making three series of tests at different values of σ, the three values of λ and hence of σ_u, σ_0, m can be obtained.

8.6 A More Rigorous Derivation of the Weibull Distribution

Weibull's original derivation of the two and three-parameter distributions was somewhat intuitive and one might ask if there are more fundamental reasons for the choice of these particular distributions. To examine this aspect, we think of a specimen as an assembly containing a large number of flaws and imagine that these flaws could be isolated and tested to determine their strength distribution $F(\sigma)$. We assume that flaws present in brittle solids do not interact and that failure occurs when the strength of a single flaw is reached. Following Weibull, we neglect any effect of compressive stress on fracture and consider only uniaxial tensile stresses. A more detailed treatment, which would include the effect of multiaxial tensile stresses, appears possible, but will not be considered here.

The probability that a flaw will have failed at a stress level $\leq \sigma$ is taken as $F(\sigma)$, and the average number of flaws per unit measure of length, area, or volume is taken as μ. In the simplest case, where the stress is constant throughout a specimen of V units (of volume, area, or length) with a total number of flaws $N = \mu V$, the probability $G(\sigma)$ that the specimen will have failed at a stress level $\leq \sigma$ is the probability that the weakest of the flaws in the specimen will have failed at a stress level $\leq \sigma$. Thus,

$$G(\sigma) = 1 - [F(\sigma)]^N = 1 - e^{N\ln[1-F(\sigma)]} \tag{8.1}$$

The number of flaws present in bulk specimens of brittle solids will be large, and as $N \to \infty$, it is necessary that $F(\sigma) \to 0$ so that $G(\sigma) \le 1$. As a result, the distribution $G(\sigma)$ may or may not exist depending on whether a limit exists for the convergence of $N\ln[1 - F(\sigma)]$. Clearly the existence of such a limit will depend on how $F(\sigma)$ approaches zero.

The distribution of the smallest values in a sample of size N drawn from a parent distribution has been treated in detail, and it has been shown that as $N \to \infty$ only three forms are possible. Two of these three forms correspond to cases in which the range of the parent distribution has no lower bound while for the third form, the range has a lower limit. In the present case, the third type is the relevant one, for the flaw strength will have a lower limit such that $F(\sigma) = 0$ for $\sigma \le \sigma_u$. Then it may be shown that provided $F(\sigma) = 0$ behaves like $C_0(\sigma - \sigma_u)^m$ as σ approaches σ_u from above, the distribution of the smallest value will always converge toward

$$G(\sigma) = 1 - e^{-NC_0(\sigma-\sigma_u)^m} = 1 - e^{-\mu V C_0(\sigma-\sigma_u)^m} \tag{8.2}$$

independent of the form of $F(\sigma)$. For example, we may assume distributions of the form

$$\begin{aligned} F(\sigma) &= 1 - e^{-C_0(\sigma-\sigma_u)^m} & \sigma > \sigma_u \\ &= 0 & \sigma \le \sigma_u \end{aligned} \tag{8.3a}$$

or

$$\begin{aligned} F(\sigma) &= 1.0 & \sigma > \sigma_u + C_0^{-1/m} \\ &= C_0(\sigma - \sigma_u)^m & \sigma_u < \sigma \le \sigma_u + C_0^{-1/m} \\ &= 0 & \sigma \le \sigma_u \end{aligned} \tag{8.3b}$$

Equation 3.2 is essentially that proposed by Weibull, with great physical insight, years before the rigorous treatments of extreme value statistics were available. Weibull expressed the distribution of strength of specimens with V units (of volume, area, or length) in the form

$$\begin{aligned} G(\sigma) &= 1 - e^{-V\left(\frac{\sigma-\sigma_u}{\sigma_0}\right)^m} & \sigma > \sigma_u \\ &= 0 & \sigma \le \sigma_u \end{aligned} \tag{8.4}$$

and estimated the parameters σ_u, σ_0, m from strength tests. Thus, the Weibull distribution is a logical choice for representing the results of strength tests on brittle solids as long as the number of flaws present is large.[3] In this distribution, the unknown flaw density μ and the constant C_0 are absorbed into a single, measurable, parameter σ_0 such that $\mu C_0 = \sigma_0^{-m}$. When the stress σ varies over the region stresses in tension, Weibull replaced the exponent in Eq. (8.4) by $-\int_V \left(\dfrac{\sigma - \sigma_u}{\sigma_0}\right)^m dV$.

Attempts have been made to deduce the entire distribution of flaw strength from mechanical tests. However, from the present discussion, it is seen that this is not possible. From results of conventional strength tests, all that is known about the distribution of flaw strength is its behavior as σ approaches σ_u from above.

8.7   Composite Using Strong Fibers and Whiskers

So far we have been discussing bulk specimens of brittle solids in which failure occurs in tension at stress levels well below the theoretical strength. However, in recent years it has been possible to produce continuous fibers and short whiskers with very high strength values. In the limit with fine fibers or whiskers, the absence or at least the reduction of flaws and dislocations may lead to an approach to the ideal strength. Griffith, who although not the first to test fine fibers, was probably the first to understand what he was doing and made the interesting comment that a single row of molecules should have either the ideal strength or none at all. In any event, typical values quoted recently for continuous fibers and short $(2\mu \times 2\mu \times 2\,mm)$ whiskers are shown in the following table.

Typical values quoted for fibers and whiskers [9]

Material	σ_{max} $(10^6\,psi)$	$E\ (10^6\,psi)$	ρ SP. GR	$\sigma_m 10^5/p$	$E 10^5/\rho$	Melt temp. $(°C)$
Whiskers (S = Sublimes)						
Graphite	2.8–3.5	98	2.2	12.7	45	3,000 (S)
Al_2O_3	2.2	76	4.0	5.5	19	2,050
Si_3N_4	2	55	3.1	6.5	18	1,900 (S)
SiC	3	100	3.2	9.4	31	2,600
Fibers						
Graphite (HiFil)	0.5	70	1.9	2.4	37	

(continued)

---

[3] Strictly speaking in using Eq. (8.4), we must assume that the number of flaws is large enough for distribution of the weakest values to converge to $G(\sigma)$ but is still small enough so that the flaws do not interact. Whether this assumption will be violated in a given case will depend on the flaw strength distribution $F(\sigma)$ and the stress level involved. For example, if $F(\sigma)$ itself should be a Weibull distribution, we do not have to specify that there should be a large number of flaws.

Typical values quoted for fibers and whiskers [9]

Material	σ_{max} (10^6 psi)	E (10^6 psi)	ρ SP. GR	$\sigma_m 10^5/\rho$	$E 10^5/\rho$	Melt temp. (°C)
Glass	0.4	10	2.5	1.6	3.9	
Nylon	0.15	0.7	1.1	1.4	0.6	
Steel	0.5	30	7.9	0.77	3.8	1,300

Currently, technical applications are based on fibers, either continuous or in lengths long compared to their diameter. Whiskers while offering great potential are difficult to handle and expensive. Generally high strength brittle fibers in a ductile metallic matrix or an elastic polymeric matrix are employed, but many other permutations have been studied. Bundles of high strength filaments present the advantage compared to monolithic solids in that the crack can be arrested by the matrix after breaking a single filament.

By contrast to natural fibers where the breaking strains are large and they can be held together by weaving and sliding friction, the high strength fibers are very susceptible to surface damage. The idealized calculations that we shall make for the strength of bundles of fibers are very sensitive to misalignment. When loading is nonuniform, the strength of bundles may be greatly decreased. The fabrication variables may be quite specific with the possibility that fabrication techniques may drop the strength well below that of an idealized bundle. With these rather gloomy warnings on the record, we will now turn to examine the behavior of a parallel bundle of elements and then consider some of the very simplest cases of composite behavior.

We consider n elements clamped at each end as in Fig. 8.13; all have the same extension and are elastic to failure. We assume that all elements have the same area but breaking loads vary. A more general treatment in which both the areas and the breaking strains vary is also possible.

First, we consider small bundles and number the breaking loads (x) of the elements in descending order of strength $x_{1(\text{strongest})}, x_2, \ldots, x_1, \ldots, x_{n(\text{weakest})}$. Say r survive out of n and load P is shared equally by survivors. Then $P \le rx_r$ and the breaking load is given by maximum value of rx_r. E.G. Daniels [3] gives test results for six threads

x_{lbs}	3.2	5.2	4.4	8.2	6.1	5.7
r	6	4	5	1	2	3
xr	19.2	20.8	20.0	8.2	12.2	17.1

Hence, if the six threads tested as a bundle, $P_{max} \le 22.0$ lbs.

Fig. 8.13 Parallel model

Note: The average thread strength is 5.46 lbs and taking $P = 6 \times 5.46 = 32.8$ lbs [very nonconservative].

Now consider many elements in parallel. For the individual elements, we can find the probability density function $f(x)$ and the cumulative distribution function $F(x)$ (Fig. 8.14). If the element is a fiber, $F(x)$ may be the Weibull distribution. Since n is large and we are numbering in order of descending strength, the survival probability can be taken as $s(x) = r/n$.

At load x in each fiber (element), the number of survivors r out of n is $r = n(s(x))$. The load carried by the bundle is $P = xr = xnS(x)$. Thus, the load-carrying capacity is given by the value of x which maximizes $xS(x)$. i.e., $P_{max} = nx_{max}S(x_{max})$ where x_{max} is found from $\frac{\delta}{\delta x}[xS(x)] = 0$. For large n, Daniels showed that if $S(x) \rightarrow 0$ more rapidly than $1/x \rightarrow 0$, then the distribution of bundle strength tends to the normal distribution with a mean load $\overline{P} = nx_{max}S(x_{max})$ and a standard deviation of bundle load $= n^{1/2}x_{max}[F(x_{max})S(x_{max})]^{1/2}$. It is often convenient to define the mean breaking load per element ignoring the fact that some elements have failed before the final failure occurs, i.e., $\overline{x}_B = \frac{\overline{P}}{n}$ with a standard deviation ψ of x_B given by

$$\psi = \frac{x_{max}}{n^{1/2}}[F(x_{max})S(x_{max})]^{1/2}$$

$$\frac{\psi}{\overline{x}_B} = \frac{x_{max}[F(x_{max})S(x_{max})]^{1/2}}{n^{1/2}x_{max}S(x_{max})} = \left[\frac{1}{n}\frac{F(x_{max})}{S(x_{max})}\right]^{1/2}$$

as $n \rightarrow \infty$, $\psi\overline{x}_B \rightarrow 0$ i.e., very reproducible values of bundle strength. A term often used is the bundle "efficiency" $\varepsilon = \overline{x}_B/\overline{x}$ where $\overline{x}$ is the average breaking load for individual elements. Say $F(x)$ is a Weibull distribution. Coleman [10] has given the following values:

m	Mean	St. Dev.
1.0	1.0	0.368
2.0	0.523	0.484
4	0.280	0.608
6	0.194	0.677
8	0.148	0.723
10	0.120	0.755
20	0.062	0.841
50	0.025	0.917
100	0.012	0.951
	0.0	1.0

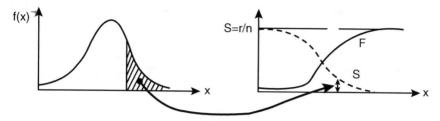

Fig. 8.14 Sketch to show $S = \int f(x)dx$

Fig. 8.15 Illustrating N
parallel elements in series

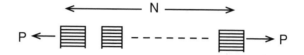

When (St. Dev./mean) < 0.25, calculations based on a normal distribution for $F(x)$ lead to about the same ε as for a Weibull distribution. If $F(x)$ is known, ψ and $\bar{x}_B$ may be calculated. Then the probability function of bundle strength is

$$k(x_B) = \frac{1}{\psi\sqrt{2\pi}} e^{-\left[\frac{(x_B - \bar{x}_B)^2}{2\psi^2}\right]}$$

and the cumulative distribution function is

$$K(x_B) = \int_0^{x_B} k(x_B)dx_B$$

Parallel Elements in Series

We consider N blocks in series as in Fig. 8.15. Each block contains many parallel elements. This situation may arise in whisker reinforced materials. The preceding discussion applied to each block. Now we ask, what is the most likely smallest value of strength in N blocks? Failure probability for each block is given by $K(x_B)$. The probability that N survive is given by $[1 - K]^N$. Hence, probability that a series of N contains at least one failure is $\theta(x_B) = 1 - [1 - K(x_B)]^N$. Most likely weakest value given by $\partial^2\theta(x_B)/\partial x_B^2 = 0$.

We quote results obtained by Gücer and Gurland [11]. For elements described by $F(x) = 1 - e^{-(x/x_0)^m}$,

	$m=2$	$m=10$
Separate elements $\bar{x}/x_0$	0.89	0.95
x_{mode}/x_0	0.71	0.95
10^5 parallel elements average strength of $\bar{x}_B/x_0$ for a layer (block)	0.71	0.79
Most probable strength of 10^5 blocks in series $\div x_0$	0.42	0.71
105 elements in series only average strength $\div x_0$	9×10^{-6}	0.1
Most probable strength $\div x_0$	7×10^{-6}	0.1

The advantage of the parallel elements model is that the dispersion in strength is greatly decreased.

It is often convenient in discussing bundles of filaments to use the normalized two-parameter Weibull distribution:

$$S = e^{-\ln\frac{1}{S'}\beta^m}$$

where β is breaking load of a filament divided by say the median or mean value and S' is the corresponding survival probability. Combining this result with $\partial/\partial x(xS(s)) = 0$, we get

$$\beta_{max} = \left(\frac{1}{m\ln\frac{1}{S'}}\right)^{1/m} \qquad \left(\because \frac{\partial}{\partial x}(\beta S(\beta))\right)$$

The fraction of filaments intact at maximum load is given by combining the expressions for β_{max} and S to obtain $S = e^{-1/m}$ at maximum load. The bundle strength can also be normalized by say the mean strength or median strength of individual filaments. Combining with the preceding equation, we find

$$\beta_{bundle} = \beta_{max}S(\beta_{max}) = \left[\frac{1}{me\left(\ln 1/S'\right)}\right]^{1/m}$$

By working with normalized quantities, we can deal with either loads or stresses. If we normalize by the mean

$$\ln\frac{1}{S'} = [\Gamma(1 + 1/m)]^m$$

$$\beta_{max} = \left(\tfrac{1}{m}\right)^{1/m}\frac{1}{\Gamma(1 + 1/m)}, \qquad \beta_{bundle} = \frac{1}{(me)^{1/m}}\frac{1}{[\Gamma(1 + 1/m)]^m}$$

If we go beyond maximum load in a stiff machine there is no catastrophic failure. Now

$$\beta_{\text{bundle}} = \beta S(\beta) < \beta_{\text{max}} S(\beta_{\text{max}})$$

The fraction R of the maximum bundle strength which is retained when this happens and a fraction S of the filaments survive is

$$R = \frac{\beta S(\beta)}{\beta_{\text{max}} S(\beta_{\text{max}})} = \frac{\beta S(\beta)}{1/[me\ln 1/S']^{1/m}}$$

But

$$\ln \frac{1}{S} = \ln \frac{1}{S'}\beta^m$$

So

$$R = S\left(me \ln \frac{1}{S}\right)^{1/m}$$

The preceding results are shown in Fig. 8.16.

We have tried to outline some of the predictions that can be made for tensile states of stress using the statistical treatment. A number of simple yet useful results can be derived for composite materials by considering the contributions of the fiber, or whiskers, and the matrix and neglecting variability in strength. A particularly clear account of this work has been given by Kelly [9].

Fig. 8.16 Strength retension ratio

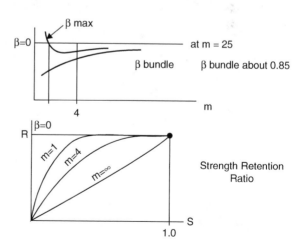

Derivation of Equations on Pages 246 and 247
p. 246

$$S(\beta) = S = e^{-\left\{\left(\ln\frac{1}{S}\right)\beta^m\right\}}$$

$$\frac{\partial}{\partial\beta}[\beta S(\beta)] = 0 \quad \text{for max load}$$

$$S(\beta) + \beta e^{-\left\{\left(\ln\frac{1}{S}\right)\beta^m\right\}}\left(-\ln\frac{1}{S}\beta^{m-1}m\right) = 0$$

Cancel $S(\beta)$ to get $1 = \beta_{\max}^m m$

$$\therefore \beta_{\max} = \frac{1}{\left(m\ln\frac{1}{S}\right)^{1/m}}$$

p. 247

Substitute $\beta_{\max}$ for β in expression for S

$$\beta_{\text{bundle}} + \beta_{\max}S(\beta_{\max}) = \frac{1}{\left(m\ln\frac{1}{S}\right)^{1/m}} e^{-\left\{\ln\frac{1}{S}\frac{1}{m}\frac{1}{\ln\frac{1}{S}}\right\}} = \frac{1}{\left(me\ln\frac{1}{S}\right)^{1/m}}$$

Using previous expression for $\beta_{\max}S(\beta_{\max})$

$$R = \frac{\beta S(\beta)}{\beta_{\max}S(\beta_{\max})} = \beta S(\beta)\left(me\ln\frac{1}{S}\right)^{1/m}$$

Let $\ln\frac{1}{S} = \frac{1}{\beta^m}\ln\frac{1}{S}$ and $S(\beta)\left(me\ln\frac{1}{S}\right)^{1/m}$

8.8   Brittle Solids Under Compressive and Multiaxial Stresses

Some years after his pioneering paper on fracture, Griffith examined the conditions for fracture under biaxial and compressive stresses. He now abandoned the energy approach and examined the local stresses at an elliptical crack. We can see two of the problems he avoided by giving up the energy approach. First, if the crack of

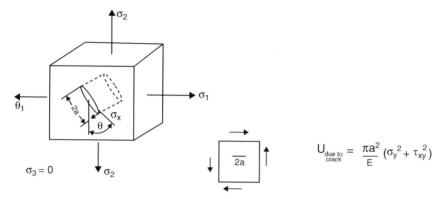

Fig. 8.17 Griffith biaxial model

Fig. 4.1 is loaded by stresses σ_y, σ_x, τ_{xy} instead of biaxial tension, the strain energy due to the crack is

$$U_{(\text{plane stress})} = \frac{\pi a^2}{E}\left(\sigma_y^2 + \tau_{xy}^2\right)$$

The stress σ_x in the direction of the major axis is not involved. The energy is independent of the sign of σ_y, but we know experimentally that brittle materials are very much stronger in compression than they are in tension. The second limitation of the energy analysis is that it assumes that cracks propagate in their own plane. Again, it is known from experiment that fracture under compressive loading may initiate somewhat away from the tip of a crack on the crack surface with propagation occurring out of the plane of the original crack.

Griffith considered the situation shown in Fig. 8.17 where sharp elliptical cracks may have any orientation θ. However, the plane of the ellipse lies in the σ_1, σ_2 plane and the length $2a$ is constant. Griffith assumed that failure would take place under multiaxial stress when the maximum tensile stress at the tip of the crack reached the same value as for fracture in uniaxial tension at a nominal stress σ_f. In this way when we compare uniaxial and multiaxial stresses, the actual stress at the tip of the crack does not appear in the final expression. On this basis Griffith found

$$\sigma_1 = \sigma_f; \theta = 0. \quad \text{if} \quad -3\sigma_1 \le \sigma_2 \le \sigma_1$$

Otherwise

$$(\sigma_1 - \sigma_2) + 8(\sigma_1 + \sigma_2)\sigma_f = 0; \quad \cos 2\theta = -\frac{1}{2}\frac{(\sigma_1 - \sigma_2)}{(\sigma_1 + \sigma_2)}$$

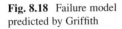
Fig. 8.18 Failure model predicted by Griffith

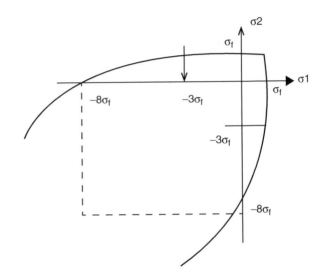

was the criterion for fracture. The result is seen more clearly if it is plotted as in Fig. 8.18. Unless the compressive stress exceeds three times the tensile, the tensile stress governs failure. For larger compressive stresses, the tensile stress at fracture becomes less and, eventually, Griffith predicts that a pure compressive stress of eight times the tensile strength will cause failure.

According to this simple model, biaxial compression would never produce failure. However, if we admit elliptical cracks with the plane of the ellipse lying in the $\sigma_1\sigma_3$ and $\sigma_2\sigma_3$ planes, then since $\sigma_3 = 0$ we would have failure if σ_1 or σ_2 reach $-8\sigma_f$. So we cut off the failure envelope of Fig. 8.18 with the dashed lines shown. Although it is sometimes stated that Griffith's prediction of the compressive strength of brittle solids is confirmed by experiment, this is by no means the case. Compression tests, as tension tests, are very difficult to carry out with precision. As secondary stresses are eliminated in the compression testing of glass, the compressive strength increases and values well over 20 times the average tensile strength are observed.

One difficulty with Griffith's derivation is that cracks may close under compressive loads and transmit both normal and shear stresses. On this basis the predictions of Fig. 8.18 have been corrected and fracture envelopes obtained for various values of the friction coefficient between the crack faces. However, a more serious limitation of the Griffith analysis for compressive stresses is that we do not have a "weakest-link" problem as in tension. Critically oriented cracks are observed to initiate, much as predicted by Griffith, and then to turn and line up with the applied compressive stress as shown in Fig. 8.19 at which point they become impotent. Very many cracks apparently have to be propagated in this manner before some critical array or echelon is formed which leads to final failure. Thus, we would not

Fig. 8.19 First crack to initiate in uniaxial compression is $\theta = 30$. After initiation, growth is as shown

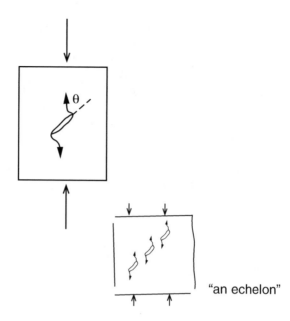

"an echelon"

expect to find as strong a size effect in compression as in tension and any comparison of the two tests has to include the specimen volume.

Weibull also generalized his approach to treat the probability of failure under multiaxial stresses. Essentially, he did this by considering the probability of failure due to the normal stress on a plane at any angle to the principal stresses. Then, integration is carried out over the range of angles for which the normal stress is tensile. Such an approach correctly predicts the weakening which is observed in biaxial tension tests relative to uniaxial strength. Weibull's approach predicts successfully the behavior of glass in the tension-compression quadrant but cannot predict failure under purely compressive states of stress.

The subject of the failure of brittle solids under compressive states of stress is one of the least well-understood areas of fracture. Despite a lot of research in recent years, most practical work is still based on the theory of Coulomb which dates back over 200 years.

Problems

1. A series of ten tensile tests on a brittle solid gave the following values:

$$6,600, \quad 8,070, \quad 8,900, \quad 9,500, \quad 9,830 \text{ psi}$$
$$10,350, \quad 10,890, \quad 11,350, \quad 11,920, \quad 12,570 \text{ psi}$$

The mean strength is 10,000 psi and the standard deviation is 1,820 psi.

Assuming that a 2-parameter Weibull distribution can be fitted to these values, estimate m in several different ways.

Can you make an estimate of the lowest strength that would be observed if 100 such specimens were tested?

If a new set of specimens is prepared with the specimen volume ten times that of the original specimens, what do you estimate as the mean strength of the larger specimens?

2. We consider tension tests on a specimen of rectangular cross-section. If the loading axis is parallel to the geometrical axis of the specimen, but eccentric by distance ε, then the specimen is subjected to a force P and a moment P^{ε}. Thus, the maximum tensile stress in the specimen is

$$\sigma_{max} = \frac{P}{bh} + \frac{6P\varepsilon}{bh^2}$$

(a) For a brittle solid containing flaws distributed through its volume in which the probability of failure of unit volume may be taken as

$$F(\sigma) = 1 - e^{-(\sigma/\sigma_0)^m},$$

you are asked to obtain an expression for the ratio:

$$\frac{\text{Median fracture load } (\varepsilon = h/6)}{\text{Median fracture load } (\varepsilon = 0)}$$

(b) Compare this result with the ratio of fracture loads (for $\varepsilon = h/6$ and $\varepsilon = 0$) which would be predicted for a brittle material with no variability in strength. Explain physically why the ratios differ.

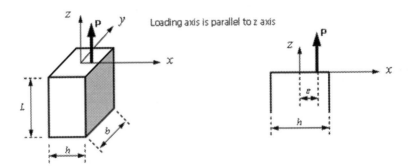

3. Consider the pure bending of a beam of rectangular cross section of a brittle solid. Assuming that fracture under tensile states of stress may be predicted using a two-parameter Weibull distribution, you are asked to obtain an

expression for the median value of moment required to fracture the beam. That is, $M_{\text{median}} = f(b, h, L, \sigma_0, m)$. Recall that the stress at a distance y from the neutral axis is given by $\sigma = \frac{My}{I}$.

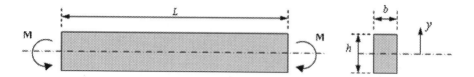

4. A frequently used design rule specifies that the strength value used in structural calculations should be taken as Mean Strength $- 3$ Standard Deviations. This rule, apparently, is based on the assumption that the strength values follow a Normal Distribution. In any event, if we are dealing with brittle solids, whose strength values follow a two-parameter Weibull Distribution say with $m = 10$, would this design rule overestimate or underestimate the strength of the weakest specimen in a batch of size N?

5. In testing short whiskers of a brittle material, the ends are embedded in a low strength matrix to which the loading is applied. As a result, the stress at each end of the whisker increases linearly over a length l from zero to the value σ_g. The stress σ_g is that in the "gauge length" L. If L is short, many whiskers will break in the grips and we do not want to discard these observations. For this reason, you are asked to derive an expression for the cumulative distribution function $G(\sigma)$ (i.e., probability that a given whisker will break at or below a gauge section stress level σ_g) which includes the probability of failure occurring in the grips. From this result can you suggest a simple expression for the effective gauge length in such a test.

   We assume that flaws are distributed along the length of these whiskers rather than volume or area and use a two-parameter Weibull distribution.

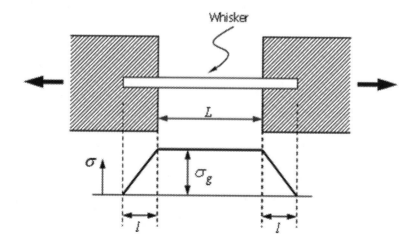

6. A bundle is made from a large number of filaments. The individual filaments show a great deal of variability and their breaking loads follow an exponential distribution. That is, the survival probability is given by $S(x) = e^{-\lambda x}$ where λ is the one parameter of the exponential distribution.

   What fraction of the filaments will survive at the maximum load-carrying capacity of the bundle?

   What is the average fracture load which would be observed in testing single filaments (expressed in terms of λ)?

7. Tests on ten glass filaments gave the following values of fracture load in pounds 6.1, 6.5, 8.5, 9.2, 11.5, 14.0, 15.0, 25.0, 30.0, 32.0 (the average is 15.8 lb).

   What would have been the maximum load-carrying capacity if these ten filaments had been tested as a bundle.

   It has been suggested that very strong bundles could be obtained by proof testing all filaments such that we discard all but the strongest 20 %.

   You are asked to predict the ratio: Strength of a bundle made from the survivors of such a proof test ÷ Strength of a bundle made from filaments that have not been proof tested. Both bundles are assumed to have the same (large) number of filaments.

8. Assuming a volume distribution of flaws, a 2-parameter Weibull distribution has been obtained for a granitic rock. The values are $m = 13$ and $\sigma_0 = 1,200$ (psi-in. units).

   You are asked to consider a test in which a cylindrical hole of radius R_j in a very large diameter cylindrical block of length L is pressurized. Assuming that the pressure P cannot penetrate any cracks present at the bore, you are asked to obtain an expression for the median pressure to produce fracture. Ignore radial and axial stresses and take the hoop stress as $\sigma_\theta = p\left(R_j/r\right)^2$. First, obtain a general result $(p \text{ median}/\sigma_0) = f(\dot{m}, R_j, L)$. Then specialize this for $m = 13$, $\sigma_0 = 1,200$, $R = 0.5$ in., $L = 10$ in.

9. Three sets of bending tests were made on specimens of a brittle solid of three different sizes. The results were:

	Tests A (15 specimens)	Tests B (4 specimens)	Tests C (4 specimens)
Strength (psi)	1,587, 1,801, 1,923 1,954, 1,623, 1,863 1,923, 2,015, 1,679 1,863, 1,953, 2,137 1,709, 1,893, 1,953	1,835, 2,163 2,190, 2,274	2,028, 2,062 2,428, 2,616
Average strength (psi)	1,858	2,116	2,283

   You are asked to use all this information to make an estimate of the Weibull parameter m for this material. One general approach for estimating m is sufficient.

References

1. Gordon JE. The new science of strong materials. Princeton University Press, New Haven, No. A920. 2006
2. Weibull W. Ingvetenskakad Handl., Stockholm, Nos. 149, 151, 153. 1939.
3. Daniels HE. The statistical theory of the strength of bundles of threads—I. Proc Roy Soc London. 1945;183A:405–35.
4. Robinson EY, Finnie I. On the statistical interpretation of laboratory tests on rock. Proc. 1969, Colloque de geotechnique, Toulouse, France. Published by I.N.S.A., Toulouse, 1971. See also UCRL reports by E.Y. Robinson.
5. Johnson LG. Theory and technique of variation research (1964), and the statistical treatment of fatigue experiments (1964). Amsterdam: Elsevier Publishing Co.; 1964.
6. Manson SS. Thermal stress and low cycle fatigue. New York: McGraw-Hill Book Co.; 1966.
7. Oh HL, Finnie I. On the location of fracture in brittle solids I—due to static loading. Int J Fract Mech. 1970;6:287–300.
8. Oh KPL, Finnie I. On the location of fracture in brittle solids I—due to wave propagation in a slender rod. Int J Fract Mech. 1970;6:333–9.
9. Kelly A. Strong solids. Oxford: Clarendon Press; 1966.
10. Coleman BD. On the strength of classical fibers and fiber bundles. J Mech Phys Sol. 1958;7:60–70.
11. Gücer DE, Gurland J. Comparison of the statistics of two failure modes. J Mech Phys Sol. 1962;10:365–73.

Printed in the United States
By Bookmasters